CW01498774

Contents

The Fred Jane Naval Wargame (1906)
including
the Royal Navy's Wargaming Rules (1921)
by John Curry

© John Curry

Fred Jane Naval Wargame first published in 1898

Published by Lulu.com 2008

ISBN 978-1-4092-4409-7

Other books by John Curry as part of the History of Wargaming Project

Donald Featherstone's War Games

Donald Featherstone's Skirmish Wargaming

Donald Featherstone's Naval Wargames

Verdy's 'Free Kriegspiel' including the Victorian Army's 1896 War Game

Paddy Griffith's Napoleonic Wargaming for Fun

See www.johncurryevents.co.uk for further details

Foreword by Richard Brooks

Fred T Jane (1865-1916) is best known as founder of *Jane's Fighting Ships*, the naval reference annual still used by navies around the world. He was also a founding father of British wargaming, the maritime counterpart of HG Wells and Robert Louis Stevenson. The *Jane Naval Wargame* and *Fighting Ships* first appeared in 1898, forming complementary halves of an integrated package. They shared a common notation for guns and armour, and used the same wooden blocks to print the game's scorer cards and ship diagrams in the book.

Jane's was not the first naval wargame. Several pioneering efforts had been demonstrated at the Royal United Services Institute since the 1870s, when steam power freed naval tacticians from dependence on the wind. Jane, however, set new standards of completeness, flexibility and realism. Much of his game equipment remains familiar: recognisable scale models, squared playing surfaces, pre-printed ship cards and turning circles. Only the strikers used for shooting need some explanation. Resembling elongated Ping-Pong bats mounting an off-centre pin beneath the head, these were held at arms length to strike at a silhouette of the target. The resulting perforations showed the impact of each shell. Only the target ship's player, however, would know exactly what damage she had suffered, if any.

The mechanism's inventor admitted its appearance was unpromising, but it did provide an objective means of comparing the vulnerability to shell fire of the different warship designs current in the 1890s. In the absence of practical experience of war at sea under steam, the *Jane Naval Wargame* was more than just a game. It had a serious professional purpose. Captain HJ May RN tested an early version while stationed at Esqimault in British Colombia, and made significant improvements. Jane then took the game onboard warships in Portsmouth Harbour. Among others, he played it with Prince Louis of Battenberg, then Flag Captain of the Channel Fleet, later Director of Naval Intelligence and First Sea Lord. In 1901 a modified version of the *Jane Naval Wargame* was issued to HM Ships, with a simplified firing system.

Captain Robert Lowry, president of the Portsmouth War College from 1907, first played Jane's game at the RUSI in 1898. Under his aegis, officers attending the War Course played wargames every Tuesday and Thursday afternoon. By then the game had evolved away from its original form, gaining in playability what it lost in late Victorian charm. It is appropriate, therefore, that John Curry has supplemented his reprint of the original *Jane Naval Wargame* with the Royal Navy's official 1921 wargame. Together these form perhaps the most positive evidence of Jane's contribution to the development of naval professionalism in the years around the Great War. John's initiative in bringing both games before the public is therefore to be applauded, and I wish him every success.

Richard Brooks, August 2008.

Introduction

Fred Jane had always been interested in naval wargaming since he played early games on the local village pond. These were probably some form of fragile model ships anchored and subject to shore-based players lobbying projectiles to simulate naval gunnery. In these games, no-one needed to consult any combat tables, as when a ship sank, it sank[1].

Over time, he developed what became a set of realistic rules. The game was initially confined to Fred Jane and his acquaintances until it received widespread publicity when it was presented in 1898 in the pages of the Engineering Journal to a wider audience.

Jane had been collecting data for his game since he was commissioned to cover a set of naval manoeuvres (i.e. war games with real ships at sea) in 1889. During this exercise, he sketched profiles of over 100 ships that he then used in his game. When he first published the game, the rules only included profiles of just four British Royal Navy ships, so there was immediately a demand for more ship profiles.

Jane initially responded by releasing data for just British and German ships, which caused outrage in the press that exclaimed that 'Germany was our friend'. In order to demonstrate impartiality, he subsequently published data for the whole world in 'All The World's Fighting Ships' (now Jane's Fighting Ships) in 1898.

This book included silhouettes of the ships and the ship cards for the original game, for example the encoding of the gun and armour data necessary for the game. For example, for Royal Navy ship Warspite, he codified it as shown below.

WARSPITE (1884).
L: 315 ft. = 96 m
Guns: 4C + 10D* + 13*.

---

[1] I have come across accounts of such games at the turn of the 19th century. Interestingly enough, there are still model ship clubs that build balsa wood ships and engage in such gunnery duels on lakes. JC.

Armour: b-c.
Sea speed: 15 kts.

The book series was successful, as it not only catered for those interested in naval matters, but it was organised to assist naval officers on the bridge of a ship in actual ship recognition while at sea. Hence, the Admiralty ordered thousands of copies per year for use on board ships and various shore establishments. The early versions of his fighting ships contained the wargaming rules, but in 1912 Messrs Sampson, Low, Marston & Co. Ltd, of London, published How to Play the Naval War Game by Fred T. Jane.

Boxed sets of ships to go with the rules were sold for those who did not want to build their own. Twelve ships and three destroyers sold for £3 3s and thirty ships and ten destroyers for £8 8s.

Fred Jane died young in Southsea, Portsmouth, England in 1916, but he had established a company that has become dominant in provided recognition guides and detailed profiles of every type of military equipment. The naval wargame had become the platform to launch what became an international company.

This edition includes

- A short biography of Fred Jane by Richard Brooks

- The original 1898 article that brought the game to wide spread public attention.

- A quick play version of the rules by Bob Cordery

- Proposed modifications to the rules

- The original 1906 version of the rules.

- The Royal Navy 1921 War Game rules that were a development of the Fred Jane game.

- The classic 1914 book on the British Navy, Your Navy as a Fighting Machine by Fred T. Jane. This largely forgotten and rare book is a fine introduction to the naval life and tactics 1900-1918.

FRED T JANE a short biography– The Man and the Wargame[2]

© **Richard Brooks (9 Dec 2007) author of** *Fred T Jane An Eccentric Visionary* published by Jane's Information Group in 1997[3].

My original title was: FIGHTING SHIPS AND INCUBATED GIRLS, reflecting the paradoxical nature of a man who was:

- founding editor of the most prestigious and longest lived naval reference work.

- artist; novelist; wargamer; pioneer motorist; and political activist.

- Portsmouth's most famous practical joker.

It is impractical to reproduce all the images used in my bibliography of Fred Jane, but I enclose a few of particular interest to wargamers, being a publicity photograph of Fred aged about 35, with his wargame, and an illustration from *Strand* magazine of May 1904, showing the Portsmouth Naval Wargame Club. The latter presents a scene familiar to modern gamers, except for the mess kit, and clouds of tobacco smoke. I also enclose an example of how *Fighting Ships* presented ship information for Admiral class battleships in 1901. This was originally engraved in a wooden block, and reused for the Jane wargame.

Fred T Jane was born 6 August 1865 in Richmond, but grew up in the West Country where his father was a clergyman. Educated at Exeter School, Fred left Devon in 1885 to make his fortune in London as a black and white illustrator for the many illustrated magazines still using line and half tone illustrations instead of photographs.

---

[2] Based in a lecture by Richard Brooks and Bob Cordery at the Conference of Wargamers, July 2007.

[3] This classic bibliography is currently unavailable. If it becomes available, it will be announced on www.johncurryevents.co.uk and through the normal book distribution channels.

He made his mark sketching the summer manoeuvres of 1890, after some years of poverty, becoming well known for his illustrations of naval events, manoeuvres, disasters, and visits by foreign ships. Some of the latter he had never seen, in particular the *Blanco Encalada*, a Chilean cruiser torpedoed in 1891. Jane's illustration of the attack was so compelling many were convinced he had been present on one of the torpedo boats, though too seasick and/or busy in the boiler room to say much about it. Only after his death was this revealed as his most long-running joke, the picture being painted in the Vicarage garden at Upottery.

Jane also produced real fiction:

- *Blake of the Rattlesnake*, in which a destroyer captain saves Britain from Franco-Russian aggression, clearly feeding off 1890s war scares.

- *The Incubated Girl*, featuring a girl hatched from an egg, and addressing the issue of vivisection (cf. HG Wells's *Island of Dr Moreau*)

- *To Venus in Five Seconds*, in which our hero is kidnapped in a space ship disguised as a summer house, and taken to Venus for medical experiments.

- *The Violet Flame*, a Wyndhamesque catastrophe in which the world is threatened by a mad scientist with a death ray, and nearly destroyed by a comet. Only the Navy stands firm to the end.

The Naval War Game and How it is Played.

The commercial failure of these and less memorable works on social themes persuaded Jane to give up fiction. It's a pity they were not more successful. He addressed similar themes to HG Wells, but was much funnier.

Meanwhile Jane was working on the naval album for which he is remembered, *All the World's Fighting Ships*, which first appeared in 1898. Unlike previous warship directories like *Brassey's Naval Annual*, it was designed for the end user: the man on the bridge trying to identify a distant silhouette on the horizon. *Fighting Ships* presented data in a revolutionary compact format. It focussed on fighting power, not gun calibres or armour thickness, whose value varied depending on when they were made. Jane classed guns and armour alphabetically, 'A' class guns being able to penetrate 'a' to 'e' class armour, 'B' class guns 'b' to 'e' armour, and so on. The format evolved rapidly as Jane absorbed criticism of early editions. Photos replaced drawings, which took too much effort to update. By 1901 the album had settled down into the form it would retain until World War II.

The wooden blocks used, as shown for the Admiral class of battleship, re-appeared in Jane's other invention, the Naval Wargame. This used:

> - 2ft squared boards representing 1 nautical mile of 2000yds

> - 1/2400 scale cork ships (1.5ins long)

> - A dice free shooting system using strikers rather like ping-pong bats to punch holes in simplified targets on flimsy paper, with a pin fixed underneath their head. There were loads of these, all with pins in slightly different positions to give an unpredictable effect. For night actions the target was covered with tissue paper)

More detailed scorers based on Fighting Ships on which the ship's owner recorded any hits. These were concealed from the firing player, who did not know what effect his fire was having. The great advantage of the firing system was that players' accuracy fell off when their own ship was hit, as it should. Hits tended to be catastrophic or irrelevant.

Like *Fighting Ships* the *Jane Naval Wargame* claimed illustrious naval patrons, including Prince Louis of Battenberg, Grand Duke Alexander of the Imperial Russian Navy, and a Captain May RN, and a close associate of the future Admiral Jellicoe. The game cost 4 guineas (£4.20), individual ships a shilling. Bits of one survive, though not the ships or boards, at Jane's Information Group. There are plans to move it to the RN Museum in Portsmouth for conservation.

Jane made enough money from these two products to pursue a rich man's hobbies:

- A battleship grey Benz racing car sounding like a destroyer, which also served as the vehicle for comic pieces in *CAR Illustrated*, in which, amongst other squibs, he compared the Sussex police to Dick Turpin.

- Standing for Parliament in 1906. His final speech ended, 'Damnation to all party politicians'. Jane's most famous political intervention was kidnapping a Labour MP called Victor Grayson outside the 1909 Party Conference, a hoax which made the front page of the national press. His most dangerous was to provoke the 'Battle of Unicorn Gate' by speaking for the Conservatives outside the dockyard.

- Boy Scout exercises involving cars, trams, and airships.

- Building a pioneer aircraft, which fell into a tree and caught fire. The inventor commented at least that was one less to include in his new book, *All the World's Airships*.

Jane's later years were less amusing. He alienated the naval establishment, by suggesting, among other things, that the RN should throw Nelson overboard. His first wife died in 1908, and his second marriage ended in separation. The First World War destroyed the information networks on which *Fighting Ships* depended. Censors prevented publication of details of British warships, which the Germans must have had already. Jane travelled the country in his open topped Benz explaining the war's progress, pouring scorn on claims that Germans were all cowards. However, a public which viewed war like a football match objected to his often paradoxical assessments. He caught a chill driving to Cheltenham, and died in March 1916, depriving us of his comments on Jutland. An unverifiable family tradition claims Jane committed suicide, depressed by the war, and the breakdown of his marriage. The death certificate gives heart failure and influenza as causes of death, so who knows.

Fred T Jane remains a paradoxical, complex figure to the end of his life and beyond. Less successful than corporate mythologists might like, his name is still associated with the original purpose of *Fighting Ships*, i.e. the provision of accurate technical information to governments, armed services and the public.

ADMIRAL CLASS.
(III.) HOWE (1885): RODNEY (1884).
10,300 tons. Complement 545.

(Dimensions) L. B. D.: 325 × 68 × 29 feet = 99 × 20·7 × 8·8 metres.

Guns:
4A (13·5 in.).
6D₀ (6 in., 26 cals., converted).
12ᵃ (6 pdr., 57 m/m).
10ᵃ (3 pdr., 47 m/m).
7 machine.
Torpedo tubes:
4 *above water.*
(*Howe* has also a bow *above water* tube).

Armour (compound):   m/m.
18″ Belt .   .   .   .   *aa* (456).
3″ Deck (steel) .   .   = *d* (76).
[Deck flat on belt.]
Protection to vitals   = *aa*.
16″ Bulkheads .   .   *aa* (405).
11½″ Barbettes .   = *aaa* (290).
12″ Barbette hoists .   .   *a* (305).
Battery bulkheads (steel).
14″ Conning tower (steel) *a* (355).

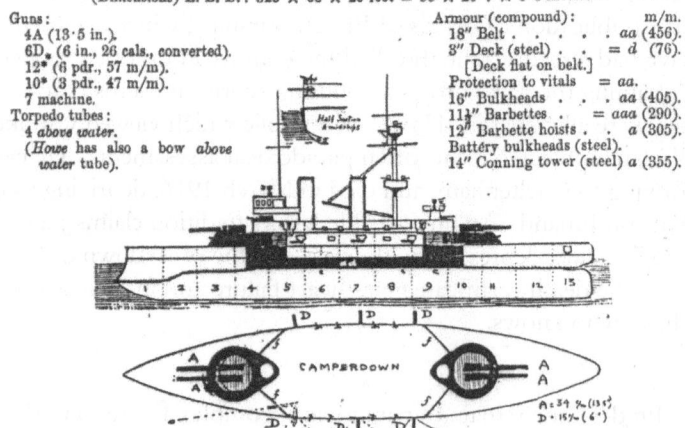

Introduction to the Fred Jane Naval Wargame

Source *The Engineer (Vol.* 86) for 9 December 1898, on p. 581. [4]

Mr Jane was not the first person to invent a naval *Kriegsspiel,* and he will not be the last. But with him lies the honour of first achieving success. Our readers already know something about his game from the various accounts of certain battles that we have from time to time published, but as we have not been, up to the present, in a position to answer the many enquiries we have had as to the manner of playing the game, we shall make no apology for describing in some detail its various parts and the method of simulating naval warfare.

In the first place, however, we must remind our readers that the naval Kriegsspiel is a game only in name. In reality it is a most instructive lesson in the capabilities of different types of ships to withstand or carry on attacks; in the practicability of evolution and of their usefulness; in the value of gun fire, and of the vulnerability of ships. In short, it puts very fairly before the players the actual problems which would face them were they commanding squadrons in times of war, and if played in seriousness cannot fail to instruct them.

The game has, we may, moreover, remind our readers, attracted the attention of many naval officers, and the rules, on which so much depends, have been, we are told, revised and approved by H.LH. Grand Duke Alexander of the Russian Navy, by Prince Louis of Battenberg, and Captain May, of our own Navy, and by Lieutenant Kawashima, of the Navy of Japan. We call attention to these facts, because it is as well to make it clear in the first place that time will not be wasted by a consideration of its main points.

The game is played upon as large a table as possible, on which are laid a number of large squares of cardboard ruled into smaller squares of about 1inch. size, the exact size does not greatly matter. The distance between the centres of any two adjacent squares represents 100 yd. On

---

[4] *I have taken the liberty of using modern standards for paragraphing to replace the 100+ sentence paragraphs of the original. This aids readability for the modern reader. JC*

this board, which is the field of operation, scale models of the ships taking part in the action are moved.

A model of the Trafalgar, which is before us, is 1.5 in. long. It is made of cork, and is suitably coloured, the guns and masts being represented by wires. The effect is much better than might be anticipated, but the value of the model lies in the fact that the exact position of the guns is shown, so that in action it is at once evident which of them will bear and on what part of the vessel attacked.

As the sides of the squares represent 100 yds, moving a model from one square to the next would be equivalent to a speed of three knots, five squares would equal 15 knots, and so on. Each move is calculated on this one-minute basis, and is not supposed to occupy a longer time. In turning, however, speed is lost. Most vessels are allowed to turn 45° on every second square, a few which do not answer well to the helm may only turn once in three, whilst others which are particularly handy may turn 45° in two adjacent squares. The turning of the model through the 45° is held equivalent to one move forward, so that if, for example, it was desired to turn a vessel of 18-knot speed, she would first move straight forward on to the third square from the one she left counting two, then would be rotated 45° counting three, then would make three moves forward to cover the distance equivalent to he manoeuvring is done in this way, which has proved sufficiently accurate and rapid for all purposes of the game.

With regard to firing and scoring, with every vessel there is supplied a large number of similar drawings of her on thin card. These drawings, elevation and plan, are divided into sections equivalent to 25 ft lengths by vertical lines ¼ in. apart, which are numbered consecutively from the bow. The armour is indicated on the a b c system, introduced, we believe by Mr Jane in his Fighting Ships. In this system the greatest protection is marked a a a a, which is proof against everything except at very close range. Then follows a a a equivalent to 30 in. of iron,

a a = 24 in. of iron; [5]

a   = 18 in.;

---

[5] Of course, Jane got this the armour coding system the wrong way around. Thin armour should have been a, the thickest armour f, so additional thicknesses of armour could be added by using g, h, i, etc... instead of resorting to aa, then aaa, then aaaa etc...

b   = 15 in.;

c   = 12 in.;

d   = 9 in.;

e   = 6 in.;

f   = thin armour.

Guns are indicated in a like manner by capital letters, but all guns over 12 in. are called A, and will, with armour-piercing shot, all penetrate a a armour at 1,000 yd, and the 110-ton gun penetrates a a a at the same distance. At 1,000 yd, the B guns, anything over 9.2 in and under 12 in. calibre, penetrate a, at 2,000 and 3,000 b, and at 4,000 and 5,000 c armour. The C D E and F guns penetrate respectively c d e and f armour at 1,000 yd. With armour-piercing shell the value of all guns is less than shot, and with common shell it is less still, but in each case the area of effect is more. Thus the 12 in. gun with armour piercing shot at 1,000 yd goes through a a armour, but its area of damage is equivalent to only 1, whilst the area of effect from a 12 in. common shell at the same range is 4, but it can only get through d armour. The amount of damage done by any shot is marked by pencil on the cards by scribbling over the number of sections destroyed. These sections are formed by the vertical 25 ft lines, and horizontal lines representing the various decks. It is presumed that the damage done is limited, upwards and downwards, by the decks.

A reduced card Sovereign[6] after she has been in action will make the manner of scoring clearer. The soft bow end, it will be noticed, has been entirely blown away up to the a a bulkhead by successive common shell; one of her masts has fallen, and one big gun aft, and two 6 in. guns on the port side have been silenced whilst one D gun in the starboard battery has been put out of action for five minutes by a shell which has passed through Section II on the port side in an oblique direction, and burst among the gun crew.

We give also a reduced target of the same ship. Many of these targets are printed on thin paper for each vessel; they are divided off, it will be observed, in a manner corresponding with the scoring card, and there are different scale drawings for 4,000, 3,000 and 2,000 yd. The firing is done with a striker, consisting of a strip of thin wood about 15 in. long, with an enlarged end, somewhere near the centre of which a short pin

---

[6] Unfortunately, the British Library were unable to find the accompanying illustrations to this article JC

point projects. The target being laid flat upon the table, one of the views, according to the range and position of the enemy, is struck at; the hole made by the pin point indicates where the shot is supposed to have hit.

From a printed card it is then ascertained, knowing the calibre of the gun fired and the nature of the projectile, how many sections are to be marked off on the scoring card. If the projectile strikes armour which it is unable to penetrate, the effect is counted nil. On the same card the rate of fire of the different sorts of guns is also marked. Thus the 12 in. gun may fire once in every two complete moves, that is, once in two minutes, whilst all guns under 9-2 may fire once a minute.

We understand, however, from Mr. Jane, that this rate of fire can with advantage be reduced, as the accuracy of the strikers is too high, and thus actions are, as a rule, over too soon. This manner of firing seems at first sight crude, and it is only after having used it for some time that its suitability is appreciated. In no two strikers is the position of the pinpoint the same, and the expanded end of the striker is of such size that a large area of the target is covered, so that a sufficiently large margin of chance prevails. It has been found that all the attempts to secure accuracy which different players adopt end, as a rule, in failure, and no matter whether a short stroke or a long stroke is made, the effect is generally the same. As a rule, moreover, the accuracy of firing gets worse as the excitement increases. It will thus be understood that the strikers do very fairly imitate the conditions of actual shooting, particularly when the rate of firing is reduced according to Mr Jane's recent suggestion. We need not go further into the manner of playing the game, which will, we hope, be sufficiently well understood from what we have already said.

There are, of course, a number of rules on which we have not touched at all, which provide for various conditions, of which it might be advisable on certain occasions to take account; but it is found that for any one action only a few of them are generally required.

It is almost out of the question to review the game critically. At first there are several points which seem as if they would gain by revision; but on a closer acquaintance one begins to see that they have their advantages, and to understand that they have not been made without great consideration and after numerous experiments. For example, it appears as if the turning rule was defective; but in actual practise it has proved itself suitable for all ordinary purposes and far simpler than various devises which suggest themselves as improvements.

We have already called attention to the strikers, and to the regulations for the number of shots per minute. By reducing the rate of fire for the big guns for all ranges, and by reducing the number of small guns fired as the range increases, the actual conditions of warfare are very well simulated, and actions take about as long as they would presumably in reality, if the crews of the various ships were equally cool and courageous. The estimates of the amount done have been worked out from official statistics, and as they have received the approval of very high authorities on the subject, we may safely conclude that they are not far from correct.

We have only touched upon the tactical game, but strategical problems can be worked out with even more value, although experts alone could hope to wrestle satisfactorily with the mass of rules and regulations which an imitation of a great strategical war must necessarily involve. There can, however, be little question that when played by persons well informed on the various points-as, for example, naval and military officers -and when well umpired, a great deal of useful information would be gained from it.

In conclusion, we must congratulate Mr Jane and those who have worked with him on what they have done, The rules alone, apart altogether from their bearing on the game, contain a mass of information upon ships of war, their armour and armament, their speeds and handiness, which cannot be found in so compact a form elsewhere, whilst a glance through those which apply particularly to the strategical game will show what a number of things have to be thought of by those who command fleets in time of war. The task Mr Jane set himself has been no light one, and great praise is due to him in that he has succeeded in grappling successfully with problems which it must have appeared at first could never be satisfactorily solved. We have little doubt the naval war game will ultimately be found at least as useful as-and it is certainly more realistic than-our military Kriegsspiels.

GERMAN SILHOUETTES.

Uniform Scale throughout, 160 feet = 1 inch,

1912-13—GERMAN DREADNOUGHTS

Kaiser
Frederich der Grosse
Kaiserin
P. Regent Luitpold
Kœnig Albert

KAISER class

1911-12—GERMAN DREADNOUGHTS

Thüringen
Helgoland
Ostfriesland
Oldenburg

HELGOLAND class (4 ships)

Fast Play Fred Jane Rules

These fast-play Fred Jane rules are included to allow the reader to start a Fred Jane naval game as quickly as possible. Some ship silhouettes (such as those supplied in the various Jane's Fighting Ships guides) and a few strikers (wooden blocks with pins in the bottom, see below for details) and battle can commence.

Rules for Tactical Naval War Games

(based on F T Jane's Naval War Game Rules of 1898 and 1905/6)

Summary by kind permission of Bob Cordery of Wargame Developments

GENERAL RULE

Nothing may be done contrary to what could or would be done in actual war.[7]

SCALES

Ship Model Scale and Sea Scale are 1:3000 *approximately*.

MOVEMENT

Each move represents one minute of time.

A speed of three knots is one square per move.

A ship may *increase* her speed by one square per move up to the maximum, and *decrease* as much as she likes subject to the rule below.

A ship wishing to *stop* travels one complete move before stopping. She can then go astern or ahead again at the normal increase rate.

TURNING

Large ships may turn 45° every second square of movement at a cost of a *temporary* loss of 3 knots of speed.

Small ships may turn 45° every square of movement at a cost of a *temporary* loss of 3 knots of speed.

A twin-screwed ship may turn 90° every square of movement at a cost of a *temporary* loss of 12 knots of speed.

---

[7] Probably the most important single rule in wargaming. All rule sets should start with it. JC.

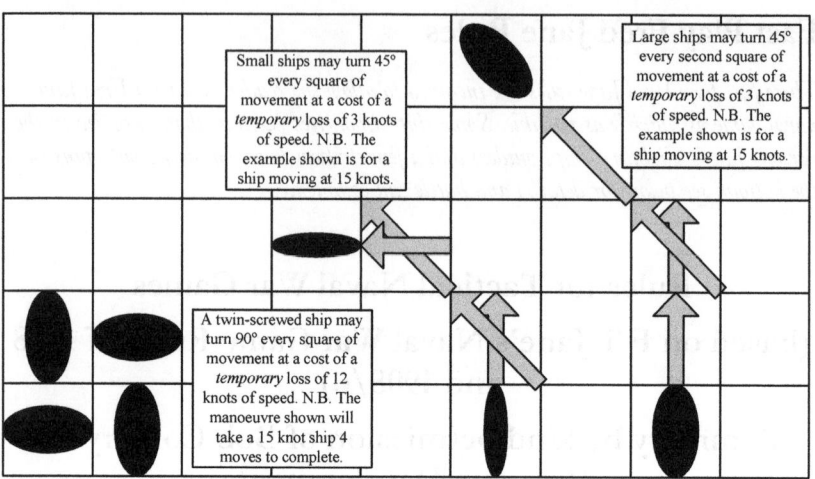

Small ships may turn 45° every square of movement at a cost of a *temporary* loss of 3 knots of speed. N.B. The example shown is for a ship moving at 15 knots.

Large ships may turn 45° every second square of movement at a cost of a *temporary* loss of 3 knots of speed. N.B. The example shown is for a ship moving at 15 knots.

A twin-screwed ship may turn 90° every square of movement at a cost of a *temporary* loss of 12 knots of speed. N.B. The manoeuvre shown will take a 15 knot ship 4 moves to complete.

TARGETS

The largest target is used for ranges up to 2000 yards, the middle target is used for ranges between 2000 yards and 4000 yards, and the smallest target is used for ranges between 4000 yards and 6000 yards.

For ranges between 6000 yards and 8000 yards the smallest target is used, but only half the guns that can bear may be struck for.

For ranges between over 8000 yards the smallest target is used, but it is covered with a piece of paper, and only half the guns that can bear may be struck for.

For hits to score they must be on the outline of the ship, and nothing is allowed as a "twixt wind and water" hit (disabling motive power, steering etc.) unless it *cuts the line* exactly between the ship and water. Anything hitting the water just below is scored as a miss.

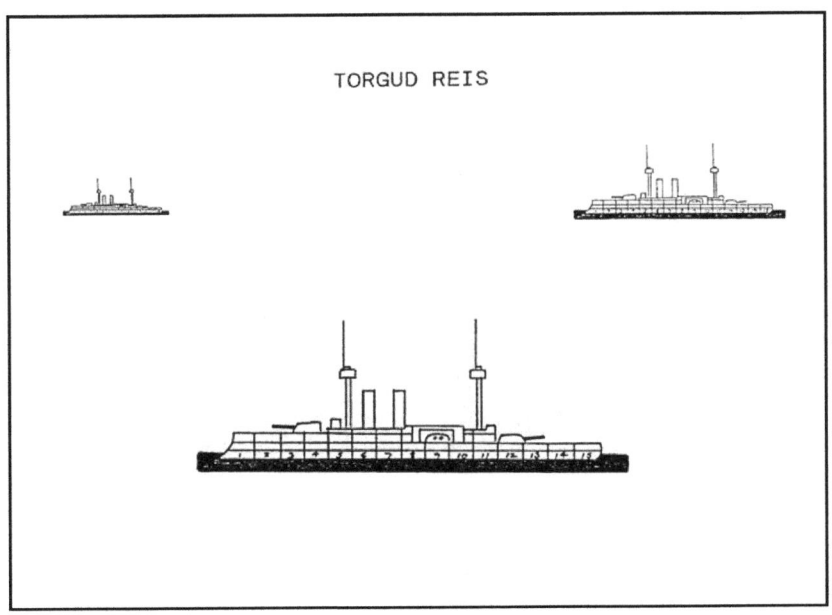

TORGUD REIS

RATES OF FIRE AND RANGES

All guns can fire once per move.

The range limits are:

      Guns of or over 9.2 inch - up to *about* 8000 yards

      Guns of or over 8.2 – 7.5 inch - up to *about* 6000 yards

      Guns of or over 6 inch - up to *about* 4000 yards

      Guns of or over 4.7 inch - up to a*bout* 3000 yards

      Guns of or over 12 pdr. - up to *about* 2000 yards

SCORERS

The printed plans of the ships are the scorers.

When a target has been fired at the scorers are to be collected in by a player on the side concerned and handed to the umpire. The same player is to collect the all scorers of his side so that the umpire can have them immediately they are required. Once damage has been assessed, the umpires will return the scorers to that player.

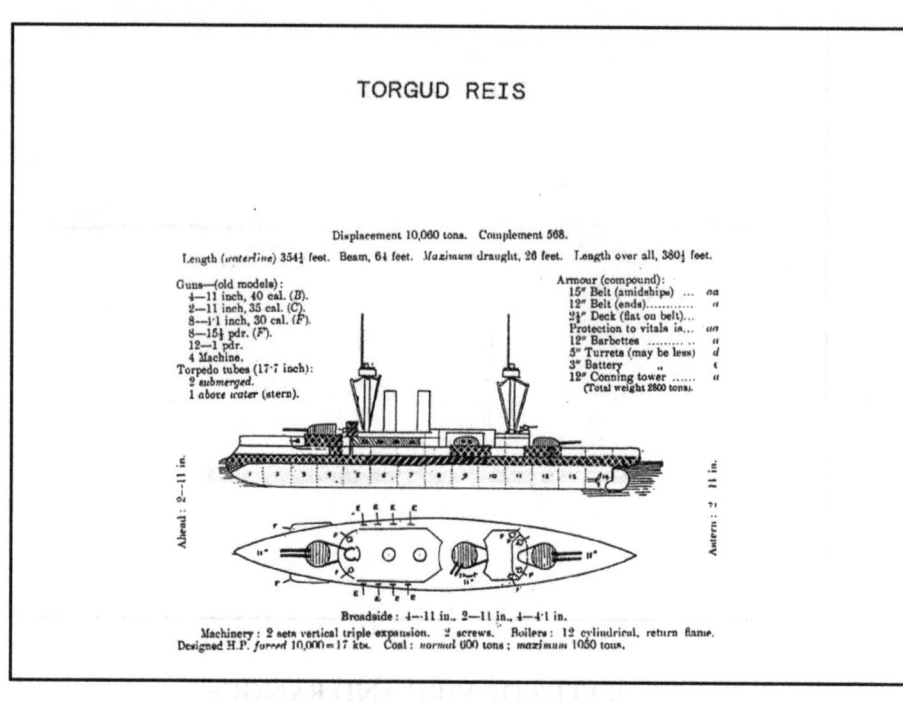

TORGUD REIS

Displacement 10,060 tons.   Complement 568.

Length (*waterline*) 354¼ feet.   Beam, 64 feet.   *Maximum* draught, 26 feet.   Length over all, 380¼ feet.

Guns—(old models):
  4—11 inch, 40 cal. (*B*).
  2—11 inch, 35 cal. (*C*).
  8—4·1 inch, 30 cal. (*F*).
  8—15¼ pdr. (*F*).
  12—1 pdr.
  4 Machine.
Torpedo tubes (17·7 inch):
  2 *submerged*.
  1 *above water* (stern).

Armour (compound):
  15″ Belt (amidships) ... *aa*
  12″ Belt (ends)............ *a*
  2¼″ Deck (flat on belt)...
  Protection to vitals is... *aa*
  12″ Barbettes .... *a*
  5″ Turrets (may be less) *d*
  3″ Battery   „   *t*
  12″ Conning tower ...... *a*
    (Total weight 2800 tons).

Abreat : 2—11 in.

Astern : ?: 11 in.

Broadside : 4—11 in., 2—11 in., 4—4·1 in.

Machinery : 2 sets vertical triple expansion.   2 screws.   Boilers : 12 cylindrical, return flame.
Designed H.P. *forced* 10,000 = 17 kts.   Coal : *normal* 600 tons ; *maximum* 1050 tons.

EFFECTS OF GUN FIRE

Essentially the best guide is the umpire's own judgement and secretion. The following rules are therefore issued merely as a guide.

	Will penetrate this grade of armour at Range		
Gun	Up to 3000 yards	Up to 5000 yards	Up to 8000 yards
AAAAA	aaaaa	aaaa	aaa
AAAA	aaaa	aaa	aa
AAA	aaa	aa	a
AA	aa	a	b
A	a	a	b
B	a	b	c
C	b	c	d
D	c	d	e
E	d	e	–
Z	e	–	–
*	–	–	–

Non-penetrating hits on belts should be marked with an "X", and a second hit in about the same spot may be scored as causing a leak by cracking the plate, with perhaps a half-a-square reduction in speed, if the first hit has been of the "very nearly through" order.

Hits on gun positions will cause the guns not to fire for *at least* one move.

Hits by shells fired from guns of or over 9.2 inch on gun muzzles will destroy that gun.

Penetrating hits on turret bases will put the turret out of action and unable to fire for D6 moves.

Penetrating hits on turrets will put the turret out of action and unable to fire for 5 or 10 moves, depending upon the toss of a coin; 5 moves if a Head is tossed, or 10 moves if a Tail is tossed.

Penetrating hits on QF batteries or casemates will destroy a gun either side of the hit.

Hits that are difficult to assess may, at the umpire's discretion, cause a reduction in a ship's rate of fire.

Hits on engines will reduce a ship's speed by *one-third* if the ship has two sets of engines and by *one-quarter* if the ship has three sets of engines.

Penetrating hits on boilers will reduce a ship's speed. The reduction will be at the umpire's discretion but a rule-of-thumb is that hits on cylindrical boilers will cause a total loss of power whereas hits on water-tube boilers will only reduce a ship's speed in proportion to the number of boilers put out of action.

Hits on a funnel base will blow out the fires concerned by the down draft created. This will cause total loss of power. If the hit is clear of the funnel base and the sides of the funnel the chance of this blowing out the fires are 1 in 12 to 1 in 18. Five hits on a funnel will cause it to fall over, reducing the ship's speed by a half-a-square per move.

Hits on unarmoured parts of the ship will destroy sections of the ship. As a general rule the number of sections destroyed depends upon the gun that has fired the shell:

12 inch	will destroy	four sections
10 inch	will destroy	three sections
9.2 inch	will destroy	two sections
8 or 7.5 inch	will destroy	one and one-half sections
6 inch	will destroy	one section
4.7 inch	will destroy	one-half sections

Penetrating hits on the water-line will cause compartments to flood. The third water-line hit may cause the ship to sink to the lower deck, and she may sink to the main deck when she has received a hit for every one thousand tons of displacement. A ship will sink when one-half of the water-line compartments have been flooded.

Penetrating hits on the water-line must be indicated by a V-shaped mark on the scorer showing which side has been hit, as repeated hits on one side may cause the ship to capsize.

Penetrating hits on the water-line will reduce a ship's speed. The reduction will be at the umpire's discretion but a rule-of-thumb is that three hits (particularly in the bow sections) will reduce a ship's speed by 3 knots (one square).

Penetrating hits on the water-line sections above the rudder may reduce a ship's manoeuvrability. The reduction will be at the umpire's discretion but a rule-of-thumb is that a hit will increase a ship's turning circle (e.g. large ships may turn 45° every third square of movement at a cost of a *temporary* loss of 3 knots of speed).

Penetrating hits on the conning tower may cause a ship to lose steering temporarily. The umpire throws a D6:

1 or 2 No effect.

3 or 4 No steering for one move. Ship remains on current course.

5 or 6 No steering for two moves. Ship remains on current course.

TORPEDOES

Torpedoes can be set for different ranges and speeds:

> Torpedoes set for 1000 yards travel, 10 squares per minute.
>
> Torpedoes set for 2000 yards travel, *say* 7 squares per minute.
>
> Torpedoes set for 3000 yards travel, 5 squares per minute.
>
> Torpedoes set for 4000 yards travel, 4 squares per minute.

A player wishing to fire a torpedo must notify the umpire that he wishes to do so at the beginning or the move or at the end. He may not do so during a move. The player must notify the umpire of the range and speed of the torpedo, and the direction in which it is to be fired.

The umpire will then keep track of the torpedo, and any ship that is in a square when the torpedo arrives is torpedoed. The ship will immediately lose 3 knots of speed and three water-line compartments will be flooded. The compartments that are flooded will be at the umpire's discretion. If a torpedoed ship has already suffered water-line hits it may capsize (see above) or sink (see above).

End of the fast play rules

Proposed Modifications to the Original Rules

The original rules require strikers made out of long pieces of wood with pins in the end. To fire, players had to select a random striker and without looking at the position of the pin, they then had to rapidly attempt to hit a paper target. The umpire then held the paper target up to the light to view the location of the hits. Each time a ship was fired on, a fresh paper target was used.

My first suggestion is to make the strikers smaller. A square block, perhaps 1 inch square, with the pin off centre is sufficient to get the same effect as the long striker. A handful of such square blocks are sufficient to prevent the players learning the location of the pin.

The second suggestion is to reuse the same paper target, rather than a fresh one for each salvo against a ship. Simply circling previous hits after each salvo allows the umpire to tell them apart. Of course, after a ship has been extensively shelled, a new target will be needed.

The original game had three silhouettes targets for each ship.

"The largest target is used for 2,000 yards, the middle sized one for 3,000 yards, the smallest for 4,000. For ranges 4,000 - 6,000 only half the guns bearing may be struck for. For ranges over that the target is covered with a piece of paper, and half strikes as for 4,000 - 6,000."

An alternative is to have only one target per ship. At less than 2,000 yards one 'strike' is made per gun firing. At 2,000 - 3,000 yards 1 'strike' per 2 shots, 3,000-4,000 yards 1 'strike' per 3 shots, 4,000-6,000 yards 1 per 4 shots, 6,000+ 1 per 5 shots. The number of shots is rounded up. This method is not exactly what Fred Jane did, but it does have the advantage of speeding up the production of the ship target cards and the actual game.

GERMAN SILHOUETTES—continued.

OLD GERMAN BATTLESHIPS

Odin   Frithjof
Aegir   Heimdal
Siegfried   Hildebrand
Beowulf   Hagen

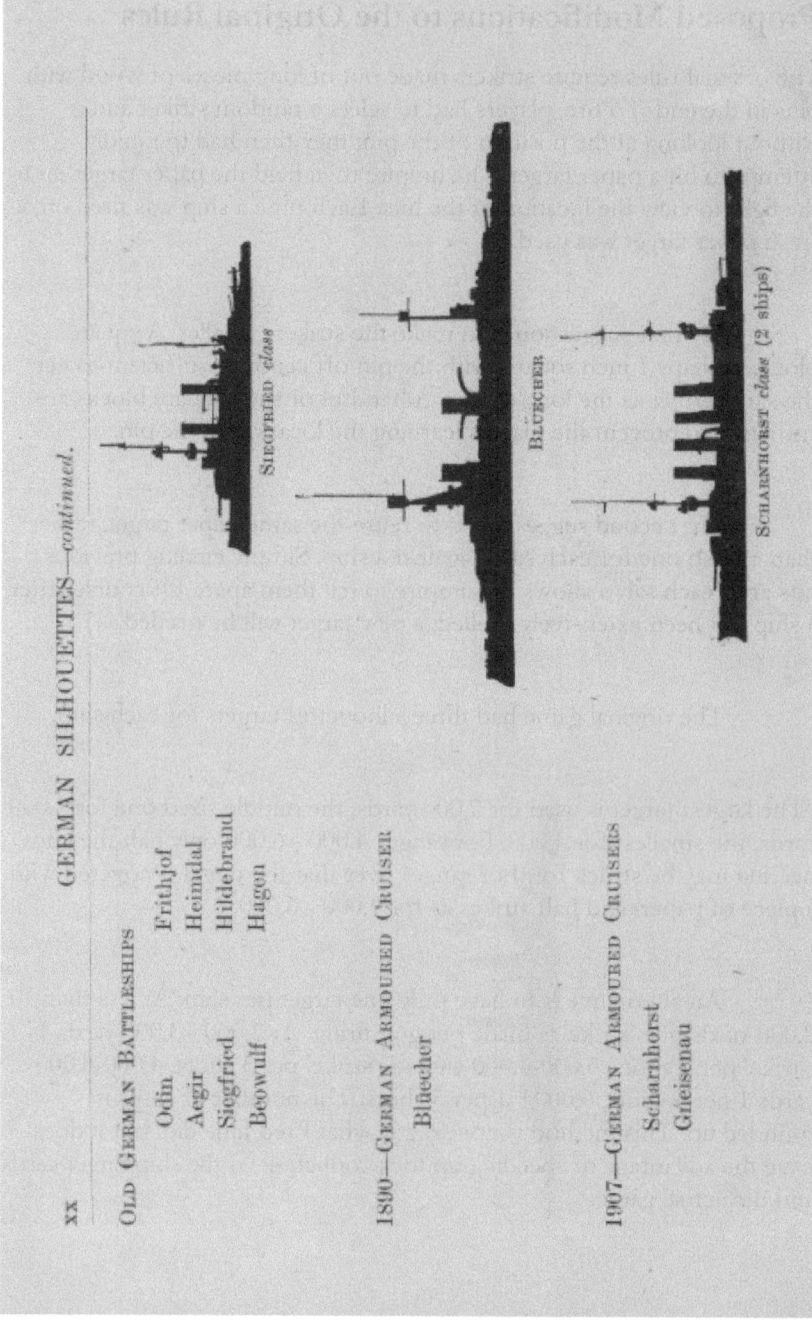

SIEGFRIED class

BLUECHER

SCHARNHORST class (2 ships)

1890—GERMAN ARMOURED CRUISER

Blücher

1907—GERMAN ARMOURED CRUISERS

Scharnhorst
Gneisenau

The Full Rules by Fred Jane

RULES
FOR
THE JANE NAVAL WAR GAME 1905/6[8]

A SEA KRIEGSPIEL SIMULATING ALL THE
MOVEMENTS AND EVOLUTIONS OF EVERY
INDIVIDUAL TYPE OF MODERN WARSHIP, AND
THE PROPORTIONATE EFFECT OF EVERY
SORT OF GUN AND PROJECTILE PART I.
TACTICAL PART II. STRATEGICAL

INVENTED BY
FRED T. JANE
AUTHOR OF "ALL THE WORLD'S FIGHTING
SHIPS," ETC., ETC.

REVISED AND APPROVED BY CAPTAIN H.I.H.
GRAND DUKE ALEXANDER MIHAILOVITCH OF
RUSSIA, I.R.N; CAPTAIN H.S.H. PRINCE LOUIS OF
BATTENBERG, R.N.; CAPTAIN H. J. MAY, R.N.;
AND LIEUTENANT R. KAWASHIMA, I.J.N.

LONDON, SAMPSON LOW, MARSTON AND COMPANY
LIMITED, ST. DUNSTAN'S HOUSE, FETTER LANE, E.C.
1898, LONDON:

PRINTED BY WILLIAM CLOWES AND SONS, LIMITED
STAMFORD STREET AND CHARING CROSS.

---

[8] There are number of versions of the Fred Jane game as his ideas developed. The first versions of the rules were originally straightforward, then suffered from added complexity, finally (to please the professional naval audience) many rules about the probability of outcome were replaced with the phrase, 'the umpire should arbitrate'. While this may have been appropriate in an age where many of the players would be familiar with such probabilities, it is frustrating from the 21st century perspective. Hence, I selected the 1905/6 version as the most playable. Other may disagree. JC.

INTRODUCTORY NOTE

The aim of this Naval War Game is to provide a thorough sea equivalent to the Army War Game. The essential idea has been to produce something by which any problems can be worked out with the greatest possible simulation of actuality, especially such as least easily lend themselves to solution on paper. In particular, the game is designed for us in all such interesting problems as those connected with "discriminating fire," and to that end not only each individual type of ship, but also every individual gun and projectile, method of protection, thickness and quality of armour is allowed for. In some of these things a certain amount of approximation has, for simplicity, been necessary, but for practical purposes the differentiation will be found sufficiently ample.

The majority of the rules do not call for special comment; all represent attempts to simulate the real thing as closely as possible; all are framed with a view to avoiding any unnecessary complications. Comparatively few of them are needed for any one game. For such as are most frequently required summaries are provided, either upon loose sheets, or upon the pieces; there is, consequently, no necessity to "learn" the game. It can be played by any naval officer after five minutes' study.

All the "shooting targets" are drawn upon the same scale, with particular attention to certain essential minutae, consequently a gun muzzle or a sighting hood has approximately its actual chance of being hit. The particular form of "striker," for localising hits, has been adopted after considerable trial with various systems. It will be found that this device affords some sort of equivalent for the moral effect of the personnel caused by damage to the matériel. It is rare to find a player "shooting" well after his ship has sustained heavy damage; very little is needed to make the aim wild. I must admit that this result is incidental rather than designed, the adoption of this device having been die to a desire to embody something that made direction easier than elevation, and did away with certainty. No tactical problem can be worked out on a basis of certainty in hitting. The accuracy obtainable is, perhaps, considerably higher than may reasonably be hoped for in action[9]; the necessity of keeping a game from taking too long to play compels that,

---

[9] From certain experiments made at target practice in board the Takasago, there is, however, reason to believe that, unless the excitement of action largely affects results, the accuracy obtainable by game methods is not much too high for modern guns.

but as a general rule accuracy is sufficiently difficult not to interfere with any evolution.

If, however, a nearer approximation to battle conditions is desired it is obtainable be reducing the rate of fire, or counting every two guns as one. It is, however, open to the objection that the chance element and the lucky shot get thereby a possibly undue value in a game that generally presupposes a comparative equality in gunnery on both sides, though it must be borne in mind that the game allows great scope for neutralising superior gunnery by superior evolutions.

A word is necessary about the gunnery rules. It is exceedingly difficult to make a general allowance for shell. Against armour they are very likely to break without bursting in the act of perforation, but the pieces carried through are likely to effect considerable damage; consequently all shell, and also shot which would carry fragments through medium armour, are given one general effect against medium armour. Another difficult problem is the behaviour of very thick Harveyised plates. There is reason to believe that if 6-inch Harvey may be considered equal to 12 inches of iron, 9 may be barely equally to 18 and 18 not equal to 36 inches of iron.

There are also innumerable side issues. This is mentioned in order to draw attention to the fact that in the letter notation an attempt is made at an approximately general allowance.

The turning and other manoeuvring qualities of ships are allowed for upon a convention. With a pair of dividers and a certain amount of patience it is of course possible to give any vessel its exact circle, but this is avoided as an unnecessary complication. In an attempt to ram, with ships of equal or nearly equal speed, it is just possible that the use of such circles might be of utility, but in such cases umpire's decision is a simple substitute. The circle selected for general use is practically the average. In the case of ships whose circle is considerably beyond the average, a perhaps somewhat unduly larger circle is given, while very handy ships have a slight undue advantage. The approximation will, however, be found satisfactory for all ordinary purposes.

Speed is also subject to arbitrary convention much as the circles are. Slight individual differences of speed do not appear to be of prime importance where the evolutions of fleets are in question: where a slightly better speed would affect results, allowance is made (see Strategical Rules). "Scoring tables" - plan and elevation with all details as to armour, etc. - of any particular ship not given with the game can be procured from the publishers, and if necessary the corresponding "shooting targets" and model ships.

None of the "shooting targets" are named. The reasons for this will be found on page 15, under the head of "General Notes."

My best thanks are due to Captain H.I.H. Grand Duke Alexander Mihailovitch of Russia, Imperial Russian Navy; Captain H. J. May, C.B., and Captain H.S.H. Prince Louis of Battenberg, K.C.B., Royal British Navy; and Lieutenant Kawashima, Imperial Japanese Navy, for looking over the proofs of the rules, and for all the kind interest they have taken in the game. I must also express my gratitude to Commander H. Russell Robinson Commander H. H. Campbell, Lieutenant Barry, and Mr. H. W. Metcalfe, all of the Royal Navy; to Mr. Kondo, of the Imperial Japanese Navy; and generally to the wardroom officers of H.M.Ss Majestic, Trafalgar, St. George (1894-8 commission), Royal Yacht, Mars, Alexandra, and Australia, and of H.I.J.M.S. Takasago, for their kindness in connection with many experimental games played on board their ships.

In conclusion it should be stated that all important shops of new type will be added to the game from time to time as they may be completed for sea, and for any new invention provision will at once be made in the rules. Any such changes will be announced in the Army and Navy gazette, Le Yacht, and one or two other papers; but, so far as possible, notice will also be sent to any shop known to be in possession of the game.

Any suggestions in the way of increased realism and simplicity, sent to me, c/o Sampson Low, Marston & Co., Ltd., Publishers, London, will be thankfully received and acknowledged. Any notes upon interesting actions will also be very welcome.

Portsmouth FRED T. JANE, June, 1898

F.T.JANE'S RULES FOR NAVAL WARGAMES (1905 - 1906 VERSION)

GENERAL RULE

Nothing may be done contrary to what could or would be done in actual war.

PROCEDURE

Admirals put down in writing all general orders etc., issued before the action. When the ships are put on the table each admiral to be seated so that his eyes are on a level with the table, and he must guess the enemy's formation as best he can in this position, and recognise the ships.

No other player may give information on any of these matters to the admirals, but in certain circumstances (when scouts are advanced, etc.) the umpire may do so at his discretion. Flag-captains will be responsible for moving the ships as directed by the admirals. They also will usually have charge of the scorer cards. One player is responsible for handing out the necessary targets to those who do the striking, also for the collection of the scorers cards as required by the umpire.

PLAYERS

A sufficient number of players will attend to the shooting. They are to aim as may have been laid down in general orders, and may fire one for each ship or group of ships, but it is better when there are many more

ships than players that some should attend to the heavy guns, others to the intermediate, leaving the light armaments (fire for which is claimed not struck for) in the hands of one special player. One player should be detailed for torpedo.

These duties can be distributed or manifolded according to the number of players available.

Should an admiral be "killed" he is to change places with his second in command.

SIGNALLING

In the early stages uninterrupted signalling may be allowed, and the admiral kept informed as to the damages received.

When closer quarters are reached the umpire will stop free signalling and subsequent messages must be written on chits, and sent through the umpire.

In order to save time there should be some recognised signal code.

A move should elapse between the making of a signal in these conditions and its reception.

The use of screens (see "moving") automatically provides for most signalling problems.

MOVES

Each move represents one minute of time. On full scale each square is 100 yards. A scale of squares should be made on a striker handle, by which all moves over the diagonals are made.

A speed of three knots is one square per move and pro rata.

On the tails of the model the maximum speed in squares is indicated.

A ship may increase her speed by one square per move up to the maximum, and decrease as much as she likes subject to the rule below.

A ship wishing to stop travels one complete move. For example a ship at 15 knots (five squares) must go on five squares. She can then go astern or ahead again at the normal increase rate.

Turning - Four types of turning-circle cards are provided - for tactical diameters of 1,000, 800, 600, and 400 yards. Each model bears the circle most akin to its real circle (see Note 1) with extreme helm. Speed loss is automatically provided for on these cards.

A ship wishing to turn with her engines may do so to the extent of turning 8 points per square, but the time occupied in turning 8 points will be the same as with the larger circles.

OPERATIONS

Before fleets move a screen should be placed between them. This prevents changes of course being replies to too quickly by the enemy. The move being made, the course or courses of the fleet should be marked in chalk on the board, and the number of the move put where the move ceases.

The screen is then removed for the rival admirals to see each other's ships. It is then replaced and the next move made, etc.

Any firing is done immediately after the screen is replaced: and the next move made while the umpire is scoring.

GUN FIRE

The targets should be marked in blue pencil to indicate capped A.P., red for H.E., and left plain for steel pointed common.

They are always to be selected for the next firing while the umpire is scoring. (See Note 2)

A player may be allowed (subject to the loss of accuracy rule) to change later to some other target when circumstances render it reasonable, but the nature of the projectile may not be changed. (Important).

TARGETS (see Note 3)

The largest target is used for 2,000 yards, the middle sized one for 3,000 yards, the smallest for 4,000. For ranges 4,000 - 6,000 only half the guns bearing may be struck for. For ranges over that the target is covered with a piece of paper, and half strikes as for 4,000 - 6,000.

Fire control - When fire is shifted from one ship to another, only half the usual number of guns may be struck for the first time the new ship is fired at.

Changing bearing - When the bearings alter rapidly (anything over 1,000 yards a minute) only half the available guns may be struck for.

Hits - Hits to score must be on the outline of the ship, and nothing is allowed as a "twixt wind and water" hit (disabling motive power, steering, etc.) unless it cuts the line exactly between ship and water. Anything hitting the water just below is scored as a miss.

When fleets move on parallel courses in the same direction, after the first fire at that range, targets to be as though the ships were a grade nearer than they are, as long as the ship fired at is the same ship. (see Note 4)

RATES OF FIRE

Guns with initial velocities over 2,700 one strike every 3 moves.

Guns with initial velocities over 2,500 one strike every 4 moves.

Guns with initial velocities over 2,200 one strike every 5 moves.

The ranges limits are :

      Guns of or over 9.2 inch up to about 8,000 yards.

      Guns of or over 8.2 - 7.5 inch up to about 6,000 yards.

      Guns of or over 6 inch up to about 4,000 yards.

      Guns of or over 4.7 inch up to about 3,000 yards.

      Guns of or over 12 pdr. up to about 2,000 yards.

As a rule no 6-inch guns are struck for, but their effects umpired. A player when sending up his target notes on it the number and calibre of secondary guns bearing, and what part of the enemy they are directed at.

The umpire marks damages at his discretion.

SCORERS

The printed plans of the ships are the scorers. All details printed upon them should be revised by the current issue of "FIGHTING SHIPS".

When the targets have been fired at they are to be collected by a player on the side concerned and handed to the umpire. The same player is to collect all the scorers of his side so that the umpire can have them immediately they are required.

The player responsible for the firing of any gun must always mark over each turret on the scorer the number of the move when it fires again. It is essential that this be done or confusion will result.

EFFECTS OF GUN FIRE

Essentially the best guide is the umpire's own judgment and discretion. The following rules are therefore issued merely as a guide.

Penetrations will be found on early page of the current "FIGHTING SHIPS" and in gun tables.

Non-penetrative hits on belt - These may conceivably do much damage, however mild their effect in experiments. A hit that does not penetrate a belt, but which goes well towards it, should be marked with a "X", and a second hit about the same spot may be scored as causing a leak, by cracking the plate with perhaps half-a-square reduction in speed if the first penetration has been of the "very nearly through" order.

Hits on gun positions - It is probable that a hit from a heavy projectile which fails to get through will be nearly as disastrous inside as one that does, on account of flying rivets, concussion, etc., etc. For every such hit the guns concerned should lose at least one fire, as, even if discharged, it is extremely improbable that they would hit anything. Shell in such cases cause no harm as a rule.

Shell-bursts on gun-muzzles - A gun actually hit by a big common shell, or by a smaller H.E. shell, would probably be destroyed, or certainly put out of action for the rest of the battle. But the chances of a fair and square hit doing this are small, and the hole made by the striker on a target is disproportionately large.

Therefore, it is incorrect to put a gun out of action for such a hit, and the best way is to throw a dice, putting a gun out of action for as many fires as the number that comes up. Thus, if a three be secured, the gun will not be allowed to struck for three fires that it would otherwise have had. (see Note 5)

For convenience, the muzzle is always taken to be the muzzle as indicated in the Targets.

Only one gun is to be so effected, and only from big shell of 9.2 inch and over.

Penetrative hits in turret bases - A dice-throw for all hits not actually in the gun-house - the gun, or guns, to be out of action for as many fires as the dice throw indicates.

Penetrative hits in turrets alongside the guns - A toss-up as to whether the guns concerned are out of action for five or ten fires.

Penetrative hits in Q.F. batteries - When small guns are affected by big penetration into batteries, etc., the case is probably reasonably met by putting out two guns for good - both sides, if the battery is penetrated, and one each side if it is the side of the deck below.

Penetrative hits on casemates - Casemates, one out each side would meet in either case.

Hits between casemates - When a big hit is secured between casemates, probably the off side will be out of action.

Hits difficult to assess - There are many hits which, though not very serious, may affect ammunition supply, etc. The umpire can meet these by reducing the rate of fire somewhat. This, indeed, is a very good method for scoring many sorts of hits.

HITS AFFECTING ENGINES, BOILERS, ETC.

Hits in engines - For ships with two sets of engines reduce speed by about one-third; ships with three engines, one-quarter.

Penetration to the boiler-rooms - In the case of cylindrical boilers, an explosion may be assumed, and all, or nearly all, speed lost. In water-tube boilered ships only a proportion of the boilers will be affected. Speed loss must depend upon the number of boiler groups.

The amount of water that enters depends upon the efficiency of the ship's complement, and this will depend upon the number of previous bad hits received.

A usual plan in umpiring is to submerge the ship to the level of the lower deck for the third penetration of the water-line, and to the main deck for a hit for each thousand tons of displacement (see "Effect of Flotation").

Obviously, a cast-iron rule is impossible here. In a general way care should be exercised not to score too heavily against motive power for any hits, for battle experience does not warrant much damage below as a rule.

Hits from Q.F., H.E. in funnels - If a 6 inch or other H.E. burst inside a funnel or funnel-base, it would probably blow out the fires concerned by the down draft created. With water-tube boilers - except, possibly, Bellvilles, - the circulation would also perhaps be affected.

The rule at present in use is as follows:-

For a fair and square hit in the funnel (i.e., clear of all boats, ship's side, etc., etc., and not touching the edge of the funnel) there are chances of the loss of all power from the boilers concerned as follows :-

1 in 12 to 1 in 18

If these odd chances are not secured then the funnel is simply holed. After about five hits of any sort on it (fair and square or not) a funnel may be assumed knocked over, but this, in a modern ship, will affect the speed very little - not more than half a square at the outside.

Some loss of accuracy of shooting may be given when between-deck hits hole the funnels, and so send smoke about the gun-decks. This is best met by putting out of action one gun per side for each funnel so hit.

HITS AFFECTING FLOTATION

As already stated, the third water-line hit may submerge the ship to the lower deck, and she may be submerged to the main deck when she has received a hit for every thousand tons of displacement.

Distributed hits are understood.

The umpire should always note, by V-shaped marks on the plan, which side hits are, as, if they are all one side, the ship may capsize.

Loss of speed for water-line hits should not be too great. The size of the ship must be taken into account.

Hits at the ends - The loss at three water-line bow sections will reduce the speed 3 knots (one square).

Raking hits - If aft the steering gear may be jammed in any ships launched before 1902 or thereabouts. The rate of fire of the big guns nearest the hit may be affected. The flotation that end will certainly be affected, and the turning circle should be increased.

CONNING TOWER HITS

It may be taken for granted that there will be other torpedo directors than those in the conning tower. The only question at issue is the condition of the steersman after a hit. This is usually umpired as follows :-

One dice throw:

> 1 or 2 - No effect.
>
> 3 and 4 - No steering for 1 move.
>
> 5 and 6 - No steering for 2 moves.

"No steering" means that the ship is to follow her original course with a certain amount of wobble, the umpire

moving her. The loss of steering for one or two moves covers the replacing of the man at the wheel, and the replacing of the commanding officer if he should be in the tower).

This applies to any shell (H.E.) of or over 7.5 inch, or any shot of or over 9.2 inch.

GENERAL SCORING

Penetrations as per "FIGHTING SHIPS".

Shell fire damage in unarmoured places :-

> 12 inch destroy four sections
>
> 10 inch destroy three sections
>
> 9.2 inch destroy two sections
>
> 8 or 7.5 inch destroy one and one-half sections
>
> 6 inch destroy one sections
>
> 4.7 inch destroy one-half section.

TORPEDO

Torpedoes set for 1,000 yards travel, 10 squares per minute.

Torpedoes set for 2,000 yards travel, say,7 squares per minute.

Torpedoes set for 3,000 yards travel, 5 squares per minute.

Torpedoes set for 4,000 yards travel, 4 squares per minute.

The range for which set must of course be decided on before the action.

Methods of firing:-

1. A player firing a torpedo notes the number and letter of the square he fires from, and also the square he aims at. For convenience he will fire only at the early or latter end of a move - each move, as it were, divided into two parts.

The number of the square fired at should be handed to the umpire or some player other than the one who fired the torpedo. Should any ship (friend or foe) be on the square named when the torpedo arrives, it is torpedoed.

Players are to make a point of honour to claim any ship (friend or foe) that passes in the track of the torpedo at any given part of the move, and which would be hit.

In order to ascertain this, it is convenient to prepare sticks marked in "square" distances. In cases of doubt toss up[10].

2. While the screen is up, anyone firing a torpedo draws a pencil line on the table from his ship towards the point aimed at; and the question of a hit, and exactly where the ship is hit is a matter of looking along the line where the torpedo has travelled. (see Note 6)

3. The simplest method of all is a dice throw 1 in 6 for a hit regardless of anything else. (see Note 7)

4. The firer notes on a piece of paper:

    (a) Course of enemy.

    (b) Range.

    (c) Enemy's estimated speed.

---

[10] i.e. spin a coin and say heads = hit, tails = miss. JC

If he has estimated these correctly, a hit is allowed.

5. Both sides to mark their courses on sectional paper, indicating the hostile ships as nearly as they can calculate from observation of the board. To fire torpedoes they lay a piece of tracing paper on their chart and on it rule the course of the torpedo, with the half moves ticked off.

To claim, they hand the tracing to the umpire who lays it on the hostile chart, and sees at a glance exactly what has happened. (see Note 8)

SUBMARINES

1. Each "2,000" yards blue square is to be lettered A, B, C, D, etc.

2. Each of the 400 little squares in the big square to be numbered 1 - 400 on each.

3. Each submarine player provides himself with a sheet of paper divided into squares, similarly lettered and numbered.

4. When he is on the surface or awash he moves on the big board with the other players, indicating his position with a small pin stuck in the board. He may take any reasonable means not to attract the enemy's attention to this pin; but he may not conceal it.

5. When submerged he retires from the table, sits with his back to it. The umpire locates his exact position for him on his own small squares, and thereafter he marks his course with a pin from that. He may not draw any pencil lines on his squares. For vision he is provided with a small fragment of looking-glass, not exceeding half an inch in diameter. From what he can see in this he must make his move, guessing distance of enemy and friend from foe as best he can. He is allowed to move half his move before the enemy moves and halt after the enemy has moved. His submerged speed will be about two squares per move, that is to say his half move will be one square. No loss for turning need be inflicted, as to allow for this his speed is always slightly below the possible. By this means unnecessary complication is avoided.

6. As he moves on numbered squares, the umpire, by looking at the big board, can always keep in mind the true relative positions of submarine and big ship. The submarine player on the other hand has to proceed by guess-work, and is extremely hampered, since his glass does not allow him to see the numbers on the big squares which the ships move on. At any time that he considers himself within torpedo range of the enemy he may name a square to the umpire in writing, and should a ship be upon that square when the torpedo is in operation a torpedo is allowed as having hit.

7. Having fired, he may not reload inside half an hour, and he must also do this on the surface or at the bottom. In no case may a submerged submarine lie by - he must be under way or he rises to the surface.

Boats with launching apparatus instead of tubes are not allowed more than one torpedo the half hour.

8. A big ship is not allowed to fire at submerged submarines, nor any chances of detection unless within one square of the submarine's course. Should a submarine submerged ever get at any time on the same square as a destroyer it is to be considered sunk or captured. In the case of picket boats even chances of the same thing.

9. A submarine on the surface is to be regarded as defenseless. Diving time varies, but in no case is there any such thing to be allowed as popping up and down again at once. No submarine in existence can yet do this.

SHIPS v. FORTS (see note 9)

Forts use special strikers without heads. When shrapnel is fired, hits, to be effective must be on the "shrapnel line" and immediately above the gun.

When the gun dot is hit at its outer edge, fire will be lost at one move - for a second hit three fires will be lost, and a third hit five fires. After that every further hit should be for eight fires, and the gun may only resume firing at half rates.

When the exact centre of the gun spot is hit, which is, of course, pure chance, ten fires are to be lost, after which the gun may resume firing at half rates. The chances of a gun being disabled by an actual hit on the muzzle are very small; a second centre hit may be allowed to do this.

A target should be reserved as a fort-scorer, and on this the P.F. positions should be marked, and placed at the umpire's disposal. By holding this, with the target to be scored from over it, to the light, he can ascertain whether a knock-out hit has been given. This will give the chances very fairly.

When it is considered that P.F. positions would be known to the enemy, they should be located at the ends of the shrapnel line, or the fort line, so that the enemy can direct fire upon the P.F. positions if they see fit.

Loss of P.F. positions should always entail use of a ship striker instead of a fort striker.

This change of striker should also always be ordered by the umpire when a fort has obviously received a demoralising fire.

The rule that the fort always has first fire at the ships, and a reply only allowed on the next move, should be enforced. Not only has a fort with modern guns a better range than a squadron, but it is certainly better fitted to pick up the enemy than ships are.

NIGHT OPERATIONS

It is always difficult to arrange these satisfactorily. There are two methods:-

Method 1. (see note 10) –

Moves are made in the dark or in reduced light, and sighting depends on actually seeing the enemy.

For searchlights, electric torches are employed, and the umpire allows them to be used while he counts three.

Special players in the light away from the board are detailed to do the firing, and these are supplied with destroyer targets, &c. At the end of each move, any ordered to do so by the players of ships they represent can fire.

For torpedoes an ordinary water squirt is used. This indicates the course of the torpedo quite clearly when the lights are turned up for inspection. (The players, of course, are moved from the table before this).

Method 2. –

Each player furnishes the umpire with his night orders, course, objectives, &c., &c., and the result is umpired.

STRATEGIC OPERATIONS

These should never be too large; it is well that small wars, or portions of wars, should always be played.

Admiralty charts are to be used.

AREA OF OPERATIONS

In addition, each Admiralty should furnish itself with a chart or tracing of each of the naval harbours allowed it, and these should be divided into 2,000 yards squares as in FIGHTING SHIPS. On these, all mine defences must be marked, any forts that are allowed, &c., &c., so that should any hostile demonstrations be attempted, the umpire has the necessary details to refer to. The number of ships that each harbour can accommodate should be noted; also any stores which, for the purposes of the "War", are limited in quantity.

The arrangement of all these details will be a matter of time, but the experience will be instructive.

SERVING OUT SHIPS, &c.

Preliminaries having been arranged, each side is given :-

(1) Its ship models.

(2) One scorer for each of its ship engaged.

(3) Targets for the enemy's ships

The models should be kept in separate boxes, each squadron by itself. Scorers should be grouped in envelopes, together with the necessary stores, &c., &c.

A cabin should be assigned to Red home bases and ports in communication, and another cabin to Blue.

When any ship or ships go to sea, the player responsible will take the models, scorers, &c., &c., and a chart into another cabin, where he will remain.

OPERATIONS BEGIN

He is to mark his course in 6 hour runs upon the Chart, as directed by the umpire. (see note 11) While in wireless distance he may communicate with his Admiralty by notes through the umpires. After that he can only communicate by going to some harbour where there is a telegraph, or by approaching a wireless or coastguard.

Detached cruisers:

Should he detach any scouts; these if they get out of wireless distance will be taken over by an auxiliary player or, should none be available, by the umpire.

Log:

He must keep a rough log of his movements and observations (see below) for inspection by umpire when required, and for forwarding to his Admiralty when opportunity offers (through umpire). All moves are dated from "War Imminent". The exact hour is fixed by the umpire. The date is known as the first day, and all time is reckoned in days from that, without regard to months or years.

SIGHTING DISTANCE

Fine day - up to 20 miles or so.

Night - from 5 miles to 1,000 yards according to weather.

SIGNALLING DISTANCES

By wireless or searchlight, about 60 miles (present allowance).

By masthead semaphore, about 12 miles, or less, according to weather.

By flags, about 5 miles, or less, according to weather.

BATTLE RANGE

Anything inside 10,000 by day; at night according to weather at umpire's discretion.

PROCEDURE

The umpire will visit each Admiralty for each 6 hour period, and then each "at sea" player in rotation. As he visits each, he will order another 6 hours. The duplicate traced courses will enable him, should ships be near each other, to ascertain the exact distances that they are apart, and whether they sight each other or not. His decision on these matters is final. At the end of each "6 hours" period, should any belligerents pass within sight of neutral coasts, or after a reasonable delay should they have gone where neutral merchant-ships would be likely to see them, he informs the players appointed to attend to neutral interests, &c. This player will then visit each "Admiralty", and give such version as he pleases of what he has been told, so covering such information as may be gleaned from the Press.

TRADE

If commerce defence enters into the scheme a special player must mark the courses of all merchant ships on a special chart in 6 hour runs. The umpire can, by noting the courses of raiding cruisers on their own charts, at once estimate captures by comparison. To capture a ship, including chasing, an average of six hours per ship is reasonable.

COAL ENDURANCE - STEAMING RULES

The following coal consumptions are usually allowed in strategic games. The figures have been arrived at after striking a mean from actual results, and are sufficiently accurate for all purposes.

Consumption in tons per 6 hours.

	Big fast and average cruisers full speed.	Small cruisers battleships, full speed.
	Tons.	Tons.
Bellville	120	75
Niclausse	128	84
Durr	132	88
Babcock	136	92
Cylindrical	140	96
Thornycroft-Schulz	144	102
Yarrow	148	106

For two-thirds speed half this consumption.

For one-third speed quarter this consumption.

The consumption of the average destroyer is placed as follows, per hour:-

Full speed   16 tons.

Four-Fifths speed   8 tons.

One-Third speed   3 tons.

The full-speed consumption is about the actual.

This works out that a destroyer can run for about 6 hours only at full speed.

Torpedo boats in proportion.

Submarines can endure from 3 to 6 hours submerged; the above-water radii vary very much. From 8 to 12 hours for a submarine, and up to about 16 for a submersible is the usual allowance: it is very fair. A petrol boat should not be allowed to increase her radius by a reduced speed on top, but lying by she of course consumes nothing.

No ship may steam at top speed consecutively for more than 6 hours without the speed being reduced about 10 per cent., and chances of break-down.

The efficiency of every ship should be reduced 10 per cent. at all speeds, and no speed above two-thirds allowed when the coal consumed passed the amount normally carried. The allowance is made for the difficulties in getting at excess coal, and for coal low in the bunkers.

When endurance is not a matter of prime importance in the special object for which the game is played the following simplification may be employed.

An imaginary amount of coal called a "unit" is introduced. Every ship burns 1 unit per 6 hours at 12 kts., 2 at 15 kts., 3 at 18 kts. and pro rata - without regard to whether she is battleship, cruiser or destroyer. With a little calculation this unit system can be adjusted to fit actual consumptions.

It is desirable to enlist an engineer officer each side who will be responsible for everything connected with the steaming of the fleet to which he is attached. Endurance is a matter of the utmost importance, and in this way much useful information will be acquired, and no impossible steaming feats attempted. No time spent upon the endurance problem is wasted.

GENERAL ITEMS

Such matters as the range of wireless, its interruption, signal stations, the maintenance of communications, cables, mine laying, counter-mining etc., etc., should be settled by general consent beforehand. The discussion of these questions is always instructive, and often leads to the formation of clear ideas on matters which before had been only very vaguely considered.

SCALE

For strategic purposes, and generally in the moves preceding and following an action in a "War", it will probably be necessary to reduce the scale of the squares. When this is done (except perhaps when half scale is used), it is very undesirable to use the proper models, as they give false ideas of ranges and distances.

The recognised scales of reduction are :-

> A. 2 minute moves half scale   = 10 squares per mile.
>
> B. 5 minute moves fifth scale   = 4 squares per mile.
>
> C. 10 minute moves tenth scale   = 2 squares per mile.
>
> D. 20 minute moves twentieth scale= 1 squares per mile.

For scales B,C, and D, no models are employed. Instead, pieces of cardboard, occupying the space occupied by the fleet on the scale employed are put down. On these, the ships are indicated by triangular pencil marks to show direction, and approximately of the relative size. (Two-thirds of a cable is the average battleship and small cruiser length, five-eights cable, big cruisers, one-third cable, destroyers). Each ship-mark should be numbered, so that when desired by the enemy her correct model can be shown, and the ship thus guessed at, or the models may be exhibited somewhere on the table in their proper order.

It is important that when reduced scales are employed, the space occupied by ships should be fairly accurate.

The following tables enable this to be secured without trouble:-

On full scale one square is 100 yards: one big blue square 1 knot.

On A scale one square is 200 yards: one big blue square 2 knots.

On B scale one square is 500 yards: one big blue square 5 knots.

On C scale one square is 1000 yards: one big blue square 10 knots.

On D scale one square is 2000 yards: one big blue square 20 knots.

The length of cables in inches is :-

	Two cables	One cable	Half cable
Full scale	5"	2.5"	1.25"
A scale	2.5"	1.25"	0.625"
B scale	1"	0.5"	0.25"
C scale	0.5"	0.25"	0.125"
D scale	0.25"	0.125"	0.0625"

Usual ranges in inches are:-

	10,000	8,000	6,000	4,000	3,000	2,000
Full scale	125"	100"	75"	50"	37.5"	25"
A scale	62.5"	50"	37.5"	25"	18.5"	12.5"
B scale	25.5"	20"	15"	10"	7.5"	5"
C scale	12"	10"	7.5"	5"	3.75"	2.5"
D scale	6.25"	5"	3.75"	2.5"	1.875"	1.25"

The approximate lengths in inches occupied by ships at two cables apart, may be put down as follows :-

			Scale	
Line ahead or abreast	A	B	C	D
2 ships	3"	1.25"	0.125"	0.3125"
3 ships	5"	2.25"	1.25"	0.5625"
4 ships	8.125"	3.25"	1.625"	0.8125"
5 ships	10.625"	4.25"	2.125"	1.0625"
6 ships	13.125"	5.25"	2.625"	1.3125"
7 ships	15.625"	6.25"	3.125"	1.5625"
8 ships	18.125"	7.25"	3.625"	1.8125"
9 ships	20.625"	8.25"	4.125"	2.0625"
10 ships	23.125"	9.25"	4.625"	2.3125"

Of these scales, D, with its twenty minutes moves, and 20 miles to each big blue square, is only required when space is very limited. C is often needed even on a big table, but as soon as possible, B should be reverted to.

It is very important, when any of the reduced scales are employed, to manage so that unsighted or unwatched ships are not seen by other players.

To avoid this, the simplest method is to have all players outside the room, the umpire calling them in turn.

Any ships not in sight of the player's vessels should be hidden from him by removal, or by paper over them.

This last is not very satisfactory: it is always better to remove altogether. If the positions of the temporarily removed vessels are marked in pencil on the board, or noted on a piece of paper by the umpire, no difficulty in re-placing will be experienced.

GENERAL NOTE ON GUN FIRE

Gun fire is arranged for upon a convention which aims at embodying the following salient features:-

a. Assuming good gunnery, sights, range-finders etc., etc., the best possible percentage of hits will decrease considerably as the range increases.

b. Guns themselves vary greatly and supposing a new and an old 12 inch to have the same rates of fire but very different velocities, it is obvious that one gun should secure many more hits than the other.

Consequently the gun-fire system is modelled upon the principle that the maximum strikes allowed represents the maximum hits obtainable in a given time at a given range with a given gun.

The misses which will be obtained in addition by the use of the "striker" represent human error, while the smaller target renders discriminating hits at long ranges much more difficult.

To the occasional difficulties of long range shooting must be added the difference between relatively unchanging and rapidly changing bearings.

Again, a ship which maintains a steady fire on one hostile ship is likely to make better shooting (other things being equal) than if she changes her target with every round or two fired.

Neither of these, nor of several other points that need not be particularised can be overlooked, if any approximation to the real thing is to be secured. It is, however, obvious that were all these features exactly embodied into the rules the resultant would be so complicated that confusion would inevitably result, to say nothing of the enormous expenditure of time that would be necessitated. Consequently a simple convention has been adopted which, while pretending to no minute accuracy will upon examination be found to afford a very fair relative approximation. (see note 12) It may be added that this system is the result of five years observation of the old system and several alternative systems that have been proposed and experimented with over long periods under all conditions. Practically it embodies something of most of the various alternatives proposed; especially those by the late Rear-Admiral H. J. May of the British Navy, and Captain Chamberlain of the U.S. Coast Artillery.

In "Hints on playing the Jane Naval War Game" some other systems were given which had advantages in certain special cases, but various considerations have led me to abolish all systems save the one given in the rules; and players are advised to adhere to this strictly as given.

Suggestions for improvements or alterations will at all times gladly be received, but it is requested that these shall be tried carefully over an extended period before submission, especially with reference to the following essential points:-

(a) Speed in playing.

(b) Ease of umpiring.

(c) Simplicity.

It may be added that all systems of points are worthless for the special purposes of the game, as these invariably degenerate into mere mathematical exercises, and of necessity ignore the "lucky shot" which must ever be a salient feature in all naval fighting. Further, they are useless for teaching guns and armour, for which the game was specially designed.

FRED T. JANE

Notes –

1) An approximation to actual circles is all that is necessary: and anything more accurate merely gives complication without any corresponding gain.

2) This represents loading with the selected projectile.

3) The thin paper diagrams are the "targets". They should be struck at gently. In order to localise hits difficult to see examine the backs of them.

4) Concentration of fire by a fleet on individual ships is too often unduly easy. An admiral may be allowed to order his squadron to concentrate on a given ship for a definite time, or concentration on a definite system; but the plan of shifting from one ship to another directly it is known that a certain amount of damage has been done, should not be allowed, because it is unreasonable to suppose that signals on such matters could be made and acted upon immediately.

The umpire should require each admiral to write down for his captains, or otherwise communicate to them before the battle, all fire control orders; and no further directions should be allowed to be given, save such as conform to the usual signalling rules of "time to take in the signal".

5) Players keep on their scorers over each turret the number of the move when it can fire again. The umpire crosses out the number and substitutes such higher number as he thinks fit.

6) This is one of the best methods; but should only be attempted by players thoroughly familiar with the game - or undue delays are likely to occur.

7) This, though crude, does not work badly, and is recommended for all games in which acquiring a knowledge of guns and armour is the special object of the game.

8) This is the best possible method, and is used when torpedo is the special object, but see note to method 2 which applies still more than this.

9) Special land blocks divided into squares for elevations, coasts and moving operations can be obtained.

They are coloured and divided into squares of varying size to indicate the differences of ground for moving over.

Infantry, &c., move one square per move; cavalry three squares. Model dockyards similarly divided are also obtainable.

10) This method is cumbersome, but it has the merit of introducing some very real problems in miniature.

11) For convenience of the umpire, the course should also be taken on tracing paper, each hour run ticked, and each 6 hours marked clearly.

12) It will be found in practice that the strikers give a substitute for "moral effect" that is to say, in nine cases out of ten, after a ship had been badly hit, her player will not shoot so well as before the hits were received.

This has frequently been noticed, and this is to be attributed to the fact that excessive anxiety and necessity to hit makes hitting more difficult than when less effort is required or made.

End of the Fred Jane Naval Wargame Rules

PLATE CXI

U.S.S. "NEW JERSEY" (Battleship).

Müller

U. S. S. NEW JERSEY

The Royal Navy Wargame, 1921[11]

For Official Use Only

Attention is called to the penalties attaching to any infraction of the Official Secrets Act

Instructions

For

TACTICAL AND STRATEGICAL EXERCISES

Carried out on Tables or Boards

January 1921

[Superseding instructions for Tactical and Strategical Exercises as carried out by the R.N. War College, Portsmouth (No 42, January, 1914).]

Admiralty

Naval Staff

Technical Section

---

[11] My thanks go to the Public Record Office staff who, through much perseverance (on their part, not mine) finally located this document. I also thank David Manley (Naval Wargames Society) for information about the rules on his website.

DREADNOUGHT (February, 1906) & Others pro.

Displacement about 18,000 tons. Complement .

Length (waterline), 520 feet. Beam, 82 feet. Maximum draught, feet.

Guns :
10—12 inch, XI. (AAAAA).
27—12 pdrs.
Torpedo tubes :
4 submerged (broadside).
1 submerged (stern).

Armour :
11″ Belt (amidships) ... aaa
″ Belt (forward) ......
″ Belt (aft) ....

Ahead:
6—12 in.

Astern :
6—12 in.

Broadside : 8—12 in.

Source JANE'S FIGHTING SHIPS 1906-07, Edited by Fred T. Jane, Published by Sampson Low, Marston, 1906. The notes say some detail may have been omitted at the request of the Royal Navy.

Tactical Investigations and Exercises Carried Out on Tables of Boards

1. **Tactical table or Board – Uses of-** the tactical table or boards is used for:-

    (a) Tactical Investigations

    (b) Tactical Exercises

2. **Object of Tactical Investigations-**

    (i) to formulate common tactical doctrines based on sound principles, which represent a consensus of informed opinion and be generally accepted as sound.

    (ii) To investigate the application of the principles of Command and Tactics to concrete problems in the organization and tactics of a fleet and in the co-operation of all arms.

    3. Conduct of a Tactical Investigation- In order to arrive at an unbiased conclusions the competitive element must be eliminated, and the plans, relative situations and moves of the opposing fleets should be freely discussed by all taking part in an investigation; both fleets being formed and moved and, if necessary, moved back and removed, as considered to be the most likely practice in accordance with the object of the investigation.

    The conclusions arrived at should be briefly summarized writing and the advantages and disadvantages of various plans formations, moves etc. under the most probable situations that may occur, should be given. Officers should thoroughly discuss the investigation before testing on the Tactical Table.

    It is important that the object of each investigation should be clearly stated as well as the principles on which each has been based.

    It is considered that from three to five officers form usually the best number for any one investigation, but where, for example, varying visibilities have to be taken into consideration, and time does not permit of the same officers considering all cases of the problem, additional sections should be formed.

4. **Object of the Tactical Exercises**

    (i) To practice and test the results obtained in tactical investigations in the most practical manner possible as a preliminary to tests at sea.

    (ii) To afford opportunity for practicing decentralization of command, and the full initiative of subordinate commanders in accordance with common doctrine and a particular plan of battle.

(iii)   To afford opportunity for exercising the practical application of tactical principles and doctrines in (a) the conduct of a fleet during the approach (b) the tactical co-operation of all units and all arms in action.

(iv)   To practice sending and receiving of enemy reports.

(v)   To exercise Staff officers in the observation of enemy movements and in keeping the strategical and tactical plots.

(vi)   To exercise officers commanding fleets and units in action on the information shown in plots.

5.   **Conduct of a Tactical Exercise-** it is considered that no hard and fast method of assessing damages should be drawn up or adhered to, but that the value of an exercise lies chiefly in drawing attention to, and, where possible, in summing up the main points of the various tactical situations presented during an exercise at the time they occur, with consequent opportunities for short discussion.

It is of little value to allow an exercise to continue or reach a conclusion if this is not done. Otherwise the situations are forgotten, and if obvious mistakes made by one or more officers taking part are allowed to pass, the results obtained may be entirely false.

A mistake that would obviously not occur in actual practice, but made owing to lack of time for preparation and dissemination of the C. in C.'s tactical ideas, or to the unreal conditions of the tactical table, should at once be pointed out and the move put back and re-moved. This, however, does not apply to mistakes which may obviously occur in actual practice due to the practical difficulties which may obviously occur in actual practice due to practical difficulties in co-operation on account of visibility, etc.

6.   **Guide to the discussion and summing up of situations presented during the exercise and of the general conduct of the exercises.** GENERAL CONSIDERATIONS.- The following general considerations, all of which have a most important influence in action, are given as a guide to appreciating the situations presented and the general conduct of the tactical exercise both by the umpires and the officers taking part. They should be fully considered before any decision is given.

I   a) System of command.

     b) Initiative of subordinate commanders.

     c) Degree of co-operation of the various units and arms.

II   d) Moral effect.

     e) Surprise; rapidity and energy of the attack.

     f) Gaining and keeping the initiative.

     g) Value of a vigorous offensive.

III   h) Concentration of superior force on a part of the enemy;
       Possibility of countering it, with method and time taken.

     i) Distribution of fire

     j) Effect of the state of sea, direction and force of wind
       position of sun, varying visibility on different bearings due to
       natural or artificial causes.

     k) Establishment of early and rapid hitting with guns and
       torpedoes.

     l) Possibility, and method of repelling torpedo attacks.

     TECHNICAL CONSIDERATIONS.   The following technical considerations, details of which are given in the Appendices, should be taken into consideration when discussing and summing up the various situations presented during the exercise:--

i) Extreme hitting ranges of various types of guns for:

     a) Indirect fire

     b) Direct fire

     c) and probable percentage hits (Appendix I)

ii) Ranges at which effective fire can be obtained from various types of guns and probable percentage hits (Appendix I).

iii) Value of varying degrees of concentration of gunfire at various ranges and probable percentage hits (Appendix I).

iv) Whether A,B, C or D arcs are bearing (Appendix VIII).

v) Torpedo fire (Appendix II).

     a) Range and speed of torpedoes.

     b) Rate of fire.

     c) Number of torpedoes fired.

     d) Ranges at which torpedoes can be fired depending on relative
     bearing and course and speed of target ships.

     e) Method and value of concentration of torpedo fire.

     f) Danger to own ships from own torpedo fire.

     g) Relative track of torpedoes relative to course of enemy.

     h) Order of enemy fleet, squadrons or divisions.

7. Duties of Officers. Officers will be selected for each fleet, Red and Blue, to carry out the following duties:-

Commander-in-Chief, with Chief Staff Officer and Staff Officers

Officers in command of Units, with Chief Staff Officer and Staff Officers

Movers.

In some cases, it may be permissible that the officer commanding a unit should also act as mover.

As a general rule, Special Ideas[12], Red and Blue, will be issued to Respective Commanders-in-Chief, who will, after consulting the officers on their side as necessary, prepare and hand in to the umpire a summary of their proposed plan of tactics. This should deal, amongst other things, with the following main points as far as applicable-

Object of the Commander-in-Chief
Main principles on which plan of battle is based.
System of command.
General battle instructions for   (1) Approach
                                    (2) Action
                                    (3) After action
Cruising order and disposition.
General dispositions for the Approach.
Guide to general dispositions during action.
Considerations given in paragraph 5 (above).
Possible enemy tactics.
Character and probable tactics of opposing Commander-in-Chief

After the exercises have commenced no further communication will be allowed between the units, except by signals passed through the umpires.

8. Method of carrying out the exercises – Officers directing the movements of the respective fleets and units will be so seated at side tables that their view of the enemy represents approximately what can be seen from the bridge of their own ships. It is a feature of the exercise that they never have anything approaching a bird's eye view of the opposing

---

[12] I,e, the aims and objectives. JC

fleet[13]. Binoculars are provided to see the ships more clearly, also charts, squared paper and instruments to enable the staffs to plot the movement and relative positions of the fleets and units as required. The umpires will at all times, kept the Commander-in-Chief and other senior officers, informed of events which they would know in practice concerning their own or enemy fleets.

The bearing and approximate range of any enemy's ships in sight may be given to any officer concerned when he asks for it. The degree of accuracy of the range given must depend on the distance and condition at the time.

REFERENCE POSITION[14]- it is essential that each fleet works on a reference position, as in actual practice. On the ships on which the referenced position is based making the signal indicating her reference position, the umpires will decide which ships fail to obtain it through being out of touch by means of visual links, and unknown to the officers concerned, will decide errors to be allotted to their supposed dead reckoning positions when making reports through the exercise.

The umpire may also unknown to the officers concerned impose on the reference position, as well as individual errors in different ships varying throughout the exercise by reason of the length of time since loosing touch after the reference position was obtained, variations of courses and speeds, and difficulties in keeping accurate D.R.[15]

REPORTING POSITIONS- Before and exercise commences, the umpires will decide to on and promulgate to all concerned, certain fixed geographical reporting positions for each side; these being similar to ZZ positions used during exercises at sea.

It will be convenient for the purpose of the exercise that one of these positions shall be the a central one, common to both sides and marked on the table or board.

Individual ships wishing to make reports should, whenever possible, obtain their positions with reference to one of these reporting positions from their own plots, but if they have been moved by the

---

[13] This effect can be achieved by seating the players on chairs and getting them to direct their ships on a table some distance away. JC

[14] The ship that everyone keeps relative to. E.g. a Captain might say sail in line astern 800 yards gap behind the lead ship i.e. the first ship should be 800 yards behind the lead ship, the second ship should be 1600 yards behind the lead ship etc. JC

[15] Direction/ Range. JC

movers in accordance with general instructions they may obtain their positions form the umpires who will make the necessary allowances for errors in reference or D.R. positions.

PLOTTING- It is essential that officers commanding fleets and units as well as ships acting as look-outs or scouts should keep strategical and tactical plots throughout the exercise. Staff officers must be detailed to the principle units for this purpose.

MOVING- the ships are moved by the movers in accordance with orders received from time to time from the officers commanding untis. The movers will be neutral, though in some cases it may be unavoidable for the officer commanding a unit to also move it. Unless otherwise directed by signal, movers will move the ships in accordance with the 'Instruction', previously issued by the Commander-in-chief or senior officer of the Unit.

LENGTH OF MOVES- The length of moves will be at the umpire's discretion[16]. Long moves may be made with the concurrence of both Commander-in-chief until in sight of the enemy. After this the moves can be six, three or two minutes, according to the circumstances.

For quick reckoning of distances run during a move, it is convenient to note that a six minutes move is one-tenth of the speed; the distance run for a three or two-minute move can then be quickly obtained by dividing the distance run in six minutes by two or three respectively[17].

USE OF SCREENS, &c- The opposing sides, and so far as possible, friendly units out of sight of each other, will be screened form one another until they actually sight. After this, as a general rules, the fleets or units are screened from one another whilst a movement is taking place. In certain circumstances, however, the move may be made with the screens removed, as it is recognized that it is otherwise difficult to carry the pictures as well as the exact bearing of the enemy in mind. The moves must then take place simultaneously, when each officer has definitively decided what he going to do.

---

[16] An early example of variable bound moves. JC

[17] Officers of this period were expected and able to carry out many mathematical calculations in their head. E.g. A problem might be, if the enemy are moving at 16 knots and we are doing 32 knots and the range is 8 nautical miles, if we launch in 3 minutes, how many seconds after launching the torpedoes, that will do 28 knots, will we know if they have missed. This would have estimated using mental arithmetic.

The position of fleets or units which are outside the visibility of the battle area should be kept plotted with other units on squared paper by the officers concerned. Umpires and moves, should as far as possible, keep the positions of these units indicated by marks on the table until they get into visibility.

SIGNALS- In order to save time, all signals will be made in plan language. They must be written on a slip, marked with War Game time of origin and passed to the movers through the umpires. A *minimum* of one to two minutes according to the circumstances must elapse between the times a manoeuvring signal is ordered to be made and carried into execution. When in single line ahead, the leader of a unit may, however, alter course in succession at any time by leading around, or alter speed without waiting to get a signal through.

SPEED- The full speeds allowed for various classes of ships are given in Appendix VIII, but are only correct for the date of issue of this publication. If the speed of columns exceed the limits given below, the ships in column must be assumed to be unable to keep exact station and their distance apart should be increased.

No. of Ships in Column	Maximum speed for column in station
5 to 8	2 knots less than available full speed of slowest ship other than the leader.
2 to 4	1 knot less that available full speed of slowest ship other than for leader.

If the slowest ship is leading, so long as she is undamaged, she may proceed at one knot less than her original full speed provided the remaining ships of her squadron are able to comply with the 'column' rule given above.

EFFECT OF SEA ON SPEED- With a sea of four of upwards before the beam, the speed of ships, especially small craft, must be reduced accordingly.

INCREASE AND DECREASE OF SPEED- in the tactical exercises, speed of ships will be altered at the following rates:-

To increase-

From below	8 knots	2½ knots per minute
	8-12 knots	1 knot per minute
	12 knots to within 3 of full speed (12-18 knots)	1 knot per minute
	Within 3 knots of full speed	1 knot in 3 minutes

To decrease

From above	12 knots (12-18 knots)	1 knot per minute
	12-8 knots	1 knot per minute

SEAGOING SPEED- Speeds within 1 knot of full speed can be maintained for four hours without penalty, after which the chances at the end of every six hours that an individual ship will fail to maintain full speed are to be:-

1 in 16 for a reduction amounting to not more than 10% of full speed.

Ships are liable to have their speed reduced by the chief umpire as he things fit, as the above changes being given as a guide for the case of a fleet.

9. Apparatus used- the movements of ships are represented by means of small model ships on tactical tables or boards which are graduated in 1 inch squares normally representing cables, but this scale can be varied if required. The scales most generally used are 10 in. = 1 mile or 10 in. = 2 miles.

The models, which represent ships about 500 ft. long, may be mounted on circular cards of 2½ diameter, so that when a division of ships on these cards is placed on the table with the cards in contact, the ships are 2½ cables apart.

The cards supplied are marked in azimuth[18] every 15 degrees. Appendix VIII gives the angular magnitudes and bearings of the arcs of fire which can bear for each class of ship.
    'A.' are includes all guns which can bear on one broadside.
    'B.' are those which bear one bow or quarter;

---

[18] In navigation, the reference plane is typically true north is considered 0° azimuth. Moving clockwise a point due east would have an azimuth of 90°, south 180°, and west 270°

'C' Ahead or stern

'D' is only used in special cases where ahead and stern fires are unequal. In such cases, C is always heavier than D.

Before using the cards for a tactical exercise, it is preferable that the limits of these areas should be marked at the edge of cards in pencil; the name or class of each ship taking part, her speed and main details of gun and torpedo armament should be also marked either on the cards or on a blackboard.

Smaller models are supplied for use as destroyers, submarines, etc. but in many cases it will be found useful to cut out flat cardboard shapes to represent the area covered by a flotilla or half flotilla in cruising or fighting formation.

For moving and altering course, protractors are provided. The straight portion of these protractors gives the distance moved in three minutes at the speed in knots corresponding to each number. For example, a ship steaming 14 knots will in three minutes move a distance on the Tactical table equal to the distance 0 to 14 on the straight edge of the protractor. The remainder of the protractor shows the track of the a ship under helm and distance travelled in three minutes at the respective speeds. Alterations of course for every 10^0 from 0^0 to 180^0 are shown.

Protractors for tactical diameters of 800 yards and 1,200 yards are provided for both the full and half scales of the Tactical board i.e.
1 in. - 200 yards.
1 in. $-$ 400 years

Measuring scales are also provided.

Some simple form of wooden or cardboard torpedo director is of use when considering the possibility of firing torpedoes and as a guide to their actual tracks.

Cruiser Problems

Cruiser problems include look-outs, scouting, supporting, screening, patrols, search, &c.

Scouting and screening exercises and various methods of patrols may be studied in the form either of 'Tactical Exercises' or 'Tactical Investigations'. These rules and instructions given in the proceeding pages are applicable to these exercises.

When cruiser exercises are carried out on the tactical table, it will usually be sufficient to plot the position of the battle fleets should they be in the area.

Cruiser problems involving search or patrols over large areas can be carried out by two different methods.

(1) When the object is to afford practice in preparing plans of search or patrol, issuing orders to the cruiser squadrons making suitable reports from the cruisers and considering their action on sighting the enemy, they may be worked out in the form of small strategical exercises. The rules and instructions given in the following pages for 'Strategical Exercises' are then applicable.

(2) If it is wished only to afford practice in planning a search or patrol dispositions, each officer or pair of officers may be given the problem to work out separately. Each officer (or pair) will work in a separate room, and will plan on tracing paper the tracks and successive positions through which his ships would pass for the whole period under consideration, supposing the enemy were not sighted. The Chief Umpire will lay down several alternative tracks for the evading ship or squadron on the chart, and each tracing will be laid over these to see in which case the plan adopted would prove successful

Strategical Exercises

The object of the strategical exercises is to practice the naval movement of two belligerent nations consequent on a situation known or partly known. Incidentally, it gives practice in writing orders and appreciations, and opportunities for studying British and foreign ships and naval resources.

Officers carrying out an exercise are supplied with the following information:-

(i) A 'General Idea' (Printed on white paper), giving the general situation up to the time at which the game begins, and say information which would be commonly known. This 'idea' is supplied to both sides.

(ii) A 'Special Idea' (printed on coloured paper) giving the positions of the particular belligerent's own fleets, and such information as would be known to him, but not his opponents; also such information concerning his enemy's ships as he would be likely to obtain through spies, newspapers, &c. This cannot always be depended upon as reliable.

To the officers taking part are apportioned duties such as:

> Admiralty
>
> Commander in Chief etc. of various fleets
>
> Officers Commanding Cruiser Squadrons
>
> Staff and the like.

Preparation of Appreciations and Plans and Orders for Strategical Exercises

APPRECIATIONS BY FLAG OFFICERS- before commencement of the exercise, the Officers detailed for the higher commands, assisted by the officers appointed as their Chief of Staff, will study the situation as set forth in the General and Special Ideas, and will study the situation as set forth in the General and Special ideas. They will prepare an Appreciation of the whole situation. This should be completed in duplicate, one copy being handed into the Chief Umpire at such time as may be ordered. This will usually be at noon on the day preceding the commencement of the exercise.

In addition to the Chiefs of Staffs, the Senior Officers may, before the exercise begins, consult any of the other officers who are told, for their respective sides, on matters on which they may have special knowledge. After the exercise commences, officers not in the same ship may meet and consult only if their ships are lying at anchor in the same harbour.

APPRECIATION[19] BY OTHER OFFICERS. On the first day of the exercises, the General and Special Ideas will be issued to all Officers, who will each prepare an Appreciation of the whole situation as it presents itself to his own side, handing it in to the Chief Umpire at such time as may be ordered. Staff Officers will not prepare separate Appreciationis, but will assist the Officers to whom they are attached.

ORDERS AND INSTRUCTIONS BY ADMIRATLIES OR SUPPERIOR COMMANDS- On the first day of the exercise, the Admiralties or other Officers in supreme command will commence the preparation of the War Plans for their own side, and at the same time, they, or the Commanders in Chief, will commence the preparation of such Operation Orders or Instructions to their subordinate commanders as may be necessary to direct the latter in making the requisite opening movements or dispositions to give effect to the plan.

These should be prepared in duplicate, one copy being handed in to the Chief Umpire.

(A sufficient number of copies of the War Plan and Operation Orders will be reproduced by the Secretariat to supply one to each officer whom they concern.)

OPERATION ORDERS BY SUBORDINATE COMMANDERS- On receiving copies of the War Plan and of the

---

[19] An appreciation is a military term for an analysis of the current situation and its military implications. E.g. the fleet has limited fuel, we need to conserve it, therefore keep the fleet in harbour and locate the enemy using scouts.

Commander-in-Chief's Operation Orders, each Officer in command of a Unit will proceed to prepare Operations Orders for the forces under his command.

First Move- When all Officers have handed in their Operations Orders, the first move will be ordered by the Chief Umpire.

STANDING ORDERS BY COMMANDER-IN-CHIEF AND OFFICERS IN COMMAND OF UNITS[20]- The Commander-in-Chief and each Officer in Command of a Unit will, as opportunity arises during the course of the exercise, prepare the following War 'Standing Orders' for his Fleet or Unit, and any other Standing Orders which may be specially called for. The preparation of these Standing Orders is not to be allowed to interfere with the progress of the exercise.

(a) *Cruising Orders under various conditions-* (when applicable)- In the case of the Commander-in-Chief this should include brief instructions to the followed by other units when cruising with the Flag. Officers of such Units should, therefore, find out the Commander-in-Chief's views before writing out their own instructions for cruising with the flag.

(b) *Fighting instructions-* To include the method of fighting his unit under various conditions. In the case of the Commander-in-Chief, this should include his proposed scheme of battle tactics for all units of the fleet likely to be present at a fleet action.

Officers in Command of Units who may be present at a fleet action should, therefore find out the Commander-in-Chiefs views on the matter before writing out instructions for their units in fleet actions.

(c) *Reliefs, Replenishment, Repairs etc* – Orders on how these matters should be prepared when applicable, if not already dealt with in the Operation Orders.

---

[20] Wargamers tend to work out tactics during a game, real navies work out their doctrine and their standard tactics before a battle.

Notes on preparation of Appreciations, Orders and Instructions

APPRECIATIONS should be of a general nature, and not over burdened with details; they should be clear and concise, and complete enough to show the reasons for the adoption of the selected object, the courses of action proposed to obtain this object, and the plan recommended to give effect to these courses.

They should be sufficiently complete to enable any other officer to write the necessary orders in accordance with the writer's intentions.

No definite rules can be laid down for writing Appreciations, but the following points will usually require consideration and the sequence in which they are places is suggested as one that will present result of a train of thought in a logical sequence.

(a) *Forces engaged-*

Our strength.

Enemy's strength

Movements of friendly forces.

Latest known movements of enemy forces

(The above should be presented so as to give the situation, both as regards relative strengths and positions, in a broad and easily grasped form. Details may well be put into tabular form as an appendix, so that the correctness of this summarised statement may be verified by the reader. It is useful to include in the tables, the bases of the forces referred to.)

(b) *What we want to do-* This must be selected with due consideration to the political consideration[21].

---

[21] This view reflected the Royal Navy view that wars are waged by politicians, but are fought by the armed forces. It is still is the Royal Navy's view.

(c) *Obstacles in our way-* Enemy's possible and probable courses of action *which may affect the attainment of our object,* and how we may overcome them.

(d) *How to how what we want.* Our possible courses of action which will achieve our objective. Our proposed course, with reasons.

(e) *Our Plan* – Our general plan, and dispositions of our fleet to carry out this course.

The detailed dispositions of the vessels comprising the several squadrons should as a rule, be left to the their own commanders.

WAR PLAN- This should comprise a brief explanation of the disposition and functions of the recipient's force.

It should include information of the disposition and functions of the other forces whose action may be connected with his own or who are operating in the same area.

It is not usually desirable to include other information that that which immediately concerns the recipient, as it not only introduce irrelevant matter, but also widens the chances of leakage where secrecy is important.

OPERATION ORDERS- The object of an Operation Order is to bring about a course of action in accordance with the intentions of the issuing Officer, suited to the situation and with full co-operation , and nothing more.

It should contain just what the recipient requires to know to carry out the operation and nothing more.

It should tell him nothing which he can and should arrange for himself.

It should avoid details, except where these are absolutely necessary.

Information regarding the enemy, or movements of one's own ships, should be strictly limited to what the recipient, or recipients, of the order require to know in carrying out the task assigned to them.

In order for an operation concerning more than one unit it will often be desirable to issue them as General Operation Orders to each unit concerned. If the issuing authority wishes to give more detailed orders to individual Officers, he should do so in the form of separate Operations Orders, which should not be included in the General Operations Orders.

In the case of detached forces not under the immediate control of the Commander-in-chief, *Instructions* for guidance may be more appropriate than actual orders, in which case fuller information as to the intention of the Commander-in-Chief and other confidential matter should be included, so as to ensure the recipient working in accordance with the spirit of the Commander-in-Chiefs intentions.

INSTRUCTINOS AND ORDERS – Operation Orders should usually contain the following subjects, but as in the case of Appreciations, it is undesirable and impossible to fix any definite rules for writing them:-

Heading   Number of order

Number of copy

Name of officer issuing order

All subordinates whom it affects

Date

List of charts and publications referred to in the body of the order

(a) The general situation given briefly including information about the enemy, and about such other portions of their own fleet as may affect the recipients of the Order.

(b) Brief summary of the intention of the Officer issuing the Order.

(c) The necessary instructions for those to whom the Order is issued, the chief essential being stated clearly in the first paragraphs, and more minute details in the later ones. Appendices should be used, if necessary, for very detailed information, and these should be referred to the body of the Order.

PATROLLING AT NIGHT- When vessels ar patrolling at night or in thick weather, without means of verifying their position they will be liable to have their estimated positions adjusted by the Chief Umpire, without being informed of the fact until afterwards.

SHIPS OR SQUADRON DETACHMENTS- When ships or squadrons are detached, the Officers Commanding must not on any account consult with any officers who are not in company with them as to what they should do. They will be given charts in separate rooms, which they must not leave, after which they can only communicate with ships and stations out of sight by means of wireless or by sending a despatch vessel. All such messages and signals must pass through the Umpire's room to be stamped and registered before they can be received and acted upon. The time which signals take to be transmitted depends on the weather, the distance apart , and possible interference.

ORDERS AND SIGNALS- All orders issued must be written out in the proper form as though they were issued by an Admiral or other Officer in Command to the ships under his orders. All signals are to be sent in code unless specifically ordered otherwise, and must be in accordance with the Signal Books.

All orders and signals should contain the following information:-

(i) Room from which issues.

(ii) Ship or fleet of origin.

(iii) Ship or fleet addressed.

(iv) Order or signal.

(v) War game date and time.

(vi) Signature of sender

The following procedure should be adopted:-

(a) When signals are sent in plain language:

At least three copies of each signal are to be made.

One copy is to be kept for reference.

Two or more copies are to be sent in, one being marked 'Chief Umpire' and the others with the name, or names of the intended recipients, one copy for each recipient.

(b) When signals are sent in code:

One copy of the code message is to be kept for reference, and one or more copies sent to the Umpire's Room, marked with the name, or names, of the intended recipients. One copy of the signification is to be sent to the Umpire's Room marked 'Chief Umpire'

Code messages and their signification should on no account be written in the same paper.

NUMBERING OF SIGNALS- All signals should be terminated with the 'time of origin'. This time should be quoted when subsequent reference is made to the signal.

SECRECY[22]- A safe custody of orders, signals, charts, &c must be attended to as in war. Such documents if not put away after working hours, and any incautious conversation carried on in public, may be made use of by the opposing side. This use, however, must be reported to, and authorised by, the Chief Umpire.

SPEED IN BAD WEATHER-　According to the wind and sea and their directions relative to the ships, care must be taken by their respective Officers working them that the speed of the ships, especially light cruisers and smaller vessels, is not allowed to exceed what is possible in actual practise, and also that if necessary they take refuge in port.

---

[22] The war game was also being used to train officers in being discreet and keeping secrets. On real operations, the expression 'careless talk cost lives' comes to mind. Jc

FUEL CONSUMPTION- The record of fuel consumption is to be kept for each ship or squadron in the book supplied for the purpose, by the Officer in Command or one of his staff.

WHEN THE ENEMY IS SIGHTED- When ships or squadrons come into contact with each other they may be placed by the Umpires on the tactical tables, when the subsequent movements will be carried out as in the Tactical Exercise.

Minor tactical situations will not, however, be worked out on the tactical tables, but will be decided by the Chief Umpire[23].

Officers in Command of squadrons or detached vessels, are to keep a short diary of events.

The Commander-in-Chief on each side, with other officers to assist him, will be required to produce afterwards a short history of the operations for record.

Appendix 1 Gunfire

1. Direct Fire.

Two gunfire tables 'A' and 'B' have been prepared.

Table A gives the rate of hitting made for the number of guns selected.

Table B shows how the gun power of a ship being hit is reduced and eventually her speed also, until she is sunk.

2. Indirect Fire. (where fittings are provided).

Insufficient data are at present available to admit of definite figures being given for the results when indirect fire is employed; but it

---

[23] A very modern wargaming concept, not to allow small actions not significant to the outcome of the campaign to take up players time.

appears probable that not more than 25% of the rate of hitting given in Table A would be obtained, if aircraft are available, otherwise nil[24].

3. Concentration of Fire.

(a) PAIRS- The fire of a pair is to be considered to be twice as effective as that of a single ship.

(b) THREES OR FOURS- this cannot well be compared with the fire of a single ship as the dispersion of the various salvos may help to produce hits (on as enemy snaking her course to avoid being hit);

As a very rough approximation it is to be considered that:

Threes are equivalent to 2.5 single ships.

Fours are the equivalent to 3 single ships.

---

[24] i.e. a spotting aircraft is needed to stand any chance of hitting a target that cannot be seen. JC

Class of ships	Max. Range Number of mins. for 1 hit on similar class of ship	Long Range Number of mins. for 1 hit on similar class of ship	Effective Range Number of mins. for 1 hit on similar class of ship	Close Range Number of mins. for 1 hit on similar class of ship	Standard number of guns bearing to produce the hits showing	Remarks
Capital Ships						
16	36,000	17,000-max 2 mins	17,000-13,000 1 min	Below 13,000 ½	8	Only 1/3 rate shown in clos 3-5 if firing v light cruiser 1/8th rate if firing v destroyer leader or destroyoer
15 in. L	23,700	16,000-max s mins	16,000-12,000 1 min	Below 12,000	8	
13.5 in. V	23,500	15,000-max 2 mins	15,000-11,000 1 min	Bellow 11,000	10	
12 in X	20,400	14,000 -max 2 mns	14,000-10,000 1 min	Below 10,000	10	

Class of ships	Max. Range Number of mins. for 1 hit on similar class of ship	Long Range Number of mins. for 1 hit on similar class of ship	Effective Range Number of mins. for 1 hit on similar class of ship	Close Range Number of mins. for 1 hit on similar class of ship	Standard number of guns bearing to produce the hits showing	Remarks
Light Division						
7.5 in. VI	20,900	12,000 – max 1 min	12,000-9,000 1 min	Below 9,000 0.4	6	Only 1/5 rate shown in clos 3-5 if firing v light cruiser
6.0 in. XII	18,800	11,000- max 1 min	11,000-8,000 1 min	Below 8,000 0.4	5	
Flotilla Leaders and destroyer leaders						Twice the rate if firing v destroyer leader or destroyer
4.7 in I	16,000	10,000 – max 6 mins	10,000-7,000 2 min	Below 7,000 0.75	5	
4 in V	13,700	9,000 – max 6 mins	9,000-6,000 2 mins	Below 6,000 0.75	4	

Secondary armaments v light cruisers	Max. Range Number of mins. for 1 hit on similar class of ship	Long Range Number of mins. for 1 hit on similar class of ship	Effective Range Number of mins. for 1 hit on similar class of ship	Close Range Number of mins. for 1 hit on similar class of ship	Standard number of guns bearing to produce the hits showing	Remarks
6 in XII	13,100	11,000- max 4 mins	11,000-8,000 1 min	Below 8,000 0.4	6	Only 1/5 rate shown in clos 3-5 if firing v light cruiser
6 in VII	12,200	11,000- max 4 mins	11,000- 8,000 1 min	Below 8,000 0.4	6	
5.5 in	17,900	11,000-max 5 mins	11,000-8,000 1 min	Below 8,000 0.4	6	Twice the rate if firing v destroyer leader or destroyer
4. in IX	13,400	10,000 max 2.5 mins	10,000-6,000 0.5 mins	Below 6,000 0.25	10	
4 in VII	11,000	9,000-max 5 mins	9,000-6,000 1.5 mins	Below 6,000 0.5	6	

Note 1.- Maximum range is given for a normal M.V.

Note 2.- Accuracy of range allowed on opening fire.

(a) If no R.F. ranges are obtained before straddling, range can be estimated within a 1,000 yards for each 10,000 yards range.

(b) If R.F. ranges are obtained before opening fire, average error is 1,000 yards in 20,000 yards.

	Range	Time to hit
No R.F. Ranges	20,000 yards	3 minutes
	15,000	2½
	10,000	2
R.F. Ranges obtained	20,000 yards	2
	15,000	1½
	10,000	1

If enemy is zigzagging, time to hit to be increased by 50%.

The timing rate for hits above therefore commences by time period given by (a) or (b).

Note 3.- Interpolation to be used with tables above as necessary.

The table above is constructed for good gunnery conditions and must be reduced by a percentage decided by a senior officer on the spot if these conditions are not obtained.

Table B

Gunfire

Class of Ship Mounting size guns	Smallest gun whose projectile can reduce offensive power of ship	Number of hits to knock out 1 gun of main armament	Remarks
Battleships			
16 in. guns		3 ½ 15 in.	
15 in. guns		3 15 in.	
14 in. guns	12	2 15 in.	
13.5 in. guns			
12 in. guns		1 15 in.	
Battle Cruiser			
16 in. guns		2.75 15 in.	
15 in. guns	12	2.5 15 in.	
13.5 in. guns		1.75 15 in.	
12 in. guns		0.75 16 in.	
Light Cruisers			Number of hit from turret guns to knock out
Large with 7.5 in.		9 6 in.	(a) 4
Medium with 6 in.	4	8 6 in.	(b) 3
Small with 4 in.		4 6 in.	(c) 2
Flotilla Leaders	4	1 6 in.	(a) 1
Destroyers		1 6 in.	(b) 1

Note 1.- 1 15 in. hit = 1.5 13.5 hits = 2 12 in hits.

Note 2.- When more than ½ the main armament guns are knocked out the speed is reduced by 10%

Note 3.- When all the main armament guns are knocked out, the ship is to be considered to have been sunk.

Note 4. – When the armour protection is of an exceptional nature, the Chief Umpire should exercise his discretion as to the application of the rules given for 'knocking out'.

C. – Rules for Effect of Gunfire on Aeroplanes

Table I

Number of rounds to be fired before aircraft can be considered out of action.

Height (Thousands yards)	Height Plane	Angle site gun	4 in.	3 in.	2 pdr	.303
10-15	80⁰	60⁰	800	1,000	-	-
	60⁰	40⁰	1,000	1,200	-	
	40⁰	20⁰	1,200	1,400	-	-
5-10	80⁰	60⁰	300	400	-	-
	60⁰	40⁰	400	600	-	-
	40⁰	20⁰	500	800	-	-
3-5	80⁰	60⁰	150	200	800	-
	60⁰	40⁰	200	250	600	-
	40⁰	20⁰	250	300	400	-
1-3	80⁰	60⁰	80	100	250	-
	60⁰	40⁰	100	120	200	-
	40⁰	20⁰	120	140	150	-
Below 1	80⁰	60⁰	30	40	100	1,000
	60⁰	40⁰	30	40	80	800
	40⁰	20⁰	30	40	60	600

Planes to be considered out of action:-

(a) One salvo of a turret guns firing shrapnel, if plane is within 1,200 yards and below 1,000 feet.

(b) Under fire from secondary armament creating 'splash' barrage at range of less than 2,000 yards for one minute.

(a) under fire of 2 pdr. or .303 as above.

II Rules for Effect of Gunfire on Airships

(a) For airships being attacked by high altitude fire from 3 in. and 4. in guns rules same as for C. Table I, only half quantities of ammunition to be taken in computing results.

(b) Any airship cruising within the following ranges to be considered out of action.

> Capital ships   5,000 yards
>
> Light cruisers   4,000 yards
>
> Other craft mounting high altitude guns 2,000 yards.

E Rules for Effects of Bombs Dropped from Aircraft

Table 1 Capital ships, Cruisers[25] and Aircraft Carriers

(a) Percentage of hits allowed when bombing from certain heights

Feet	Percentage of hits
100	100
500	60
1,000	35
2,000	20
3,000	15
4,000	10
5,000	7
6,000	5
7,000	3
8,000	2
9,000	Nil

(b) Effects of Hits

Weight of Bomb	Number of hits required to put ½ personnel not under cover out of action	Number of hits to put one gun of secondary armament out of action[1]
20	6	4
100	4	3
220	2	2
300	1	1

Note [1] This casualty only to be imposed in capital ships where the secondary armament guns are in an exposed position on the upper deck e.g. 'Hood'

---

[25] e.g. Courageous and Glorious

Table 2 Light Cruisers

(a) Percentage of hits allowed when bombing from certain heights

Feet	Percentage of hits
100	100
500	40
1,000	25
2,000	15
3,000	10
4,000	5
5,000	2
6,000	Nil

(b) Effects of Hits

Weight of Bomb	Number of hits required to put ½ personnel not under cover out of action	Number of hits to put one gun of secondary armament out of action[1]
20	6	4
100	4	3
220	2	2
300	1	1

Table 3 Flotilla Leaders, Destroyers and Submarines

(a) Percentage of hits allowed when bombing from certain heights

Feet	Percentage of hits
100	100
200	50
500	35
1,000	20
2,000	10
3,000	5
4,000	Nil

(b) Effects of Hits

Weight of Bomb	Number of hits to put ship out of action[1]
20	4
100	3
220	2
300	1

Table 4 Trawlers, Patrol Craft, P Boats CMR's

(a) Percentage of hits allowed when bombing from certain heights

Feet	Percentage of hits
50	100
100	75
200	50
500	25
1,000	10
2,000	5
3,000	Nil

(b) Effects of Hits

Weight of Bomb	Number of hits to put ship out of action[1]
20 -30	1

F.- Rules for Effect of Machine-gun Fire from Aircraft

(a) In capital ships, cruisers, light cruisers, aircraft carriers, torpedo craft and patrol vessels.

1/8th of personnel in exposed positions[26] to be considered out of action if plane passes over ship at height less that 200 ft and in such a way to bring machine guns into play.

(b) C.M.R.'s To be considered out of action, if under fire from aircraft at range not exceeding 1,000 yards for one minute.

---

[26] remember on light ships, such as destroyers, the Captain ran the ship from an open bridge, to maximize visibility. JC

Appendix 2 Torpedo Firing

Ranges and Speeds of Torpedoes- Torpedoes are allowed the speeds and ranges given below.

18 inch	Yards	Approx Distance Run per Minute[27]	Reference Letter for torpedo or Setting
35 knots	3,000	1,100	(a)
29	7,000	950	(b)
35	2,500	1,100	(c)
29	4,000	950	(d)
21 inch			
35	8,000	1,100	(e)
29	12,000	950	(f)
25	15,000	850	(g)

Allocation of Torpedoes- the above torpedoes are allocated to British ships and torpedo craft as given below. Foreign ships will be credited with torpedoes similar to existing British types.

Class of Ship	Torpedo or Setting
18 inch Tube Battleships and Battle Cruisers	(a) and (b)
All other surface craft except C.M.R..s	(c), (f) and (g)
C.M.R.s and Airplanes	(c) and (d)
J and K Class Submarines	(a) and (b)
All other submarines	(e), (f) and (g)

---

[27] Approximate postion of running torpedoes at the end of each minutes by be found by means of range measuring scale

Number of Torpedoes allocated to Ships[28] Ships may be allowed to carry 6 torpedoes per submerged tube. Light cruisers, 6 per A.W. broadside. Destroyers, 6 per ship. Submarines, 2 per tube.

Rate of fire Torpedoes may be fired from submerged tubes at the rate of 1 per tube every 2 minutes. A.W. tubes, one broadside in 20 seconds.

Method of firing Torpedoes will be fired as opportunity offers, unless orders to the contrary are given.

The mover will make an arrow or dot on the course of the ship firing at the point where the torpedo was discharged, noting the time.

Long Range Firing In the case of the long range adjustment, when sufficient time has elapsed for the torpedo to the reach the enemy, a line is drawn from the arrow or slot in the direction in which the torpedo runs. If the path of the torpedo passes through the enemy line, the shot is allowed and chances of hitting are given as shown in table below.

Table for Changes for Long Range Firing at Enemy Line Allowing for Errors of Torpedoes

Track Angle	45^0	90^0
Enemy Ships 5 cables[29] apart	$1/10^{th}$- $1/20^{th}$	$1/5^{th}$ – $1/10^{th}$
3 cables apart	$1/6^{th}$- $1/12^{th}$	$1/3^{rd}$- $1/6^{th}$

Short Range Firing In the case of short range adjustment, when sufficient time has elapsed for the torpedo to reach the enemy, the shot may be claimed. A line is drawn form the arrow or dot to the position of the ship fired at.

If the torpedo was fired to run within 10^0 of the line the shot is allowed and chances of hitting given from table below.

---

[28]. See also Appendix VIII

[29] cable is $1/10^{th}$ of nautical mile, 182 meters, 608 feet. JC

Table of Changes for Torpedoes are Set for Short Range
35 Knots at Individual Ships[30]

Range	Angle between track of Torpedo and Course of Enemy		
Length of run	Up to 30^0	45^0	90^0
8,000-5,000	Nil	1/12th	1/9th
3,000-5,000	Nil	1/8th	1/6th
2,000-3,000	Nil	1/5th	1/4th
2,000-1,000	Nil	1/4th	1/3rd
1,000-500	Nil	2/3rd	1/2

Note ½ the above chances are allowed at night.

Assessment of Damage – Should a torpedo strike a ship of not less than 5,000 tons, she can proceed at 2/3rds her original speed for 6 hours and afterwards at half her original speed.

If a ship of over 3,000 tons is hit by two torpedoes she is reduced to 6 knots. The same applies to a ship of 5,000 tones or less which is hit by one torpedo.

---

[30] At discretion of Chief Umpire, depending on the position in enemy's line torpedo passes through and the number of ships in the line.

Appendix III Wireless and Telegraphy and Telegraph Cables[31]

Wireless Telegraphy For details see Appendix B to British WT Instructions. Also Telegraph Chart of the World (3 sheets).

Time Signals take to get through The umpires will decided when a message gets through and will send it to the officer concerned, marked with the time of supposed receipt. If the signals are of a lengthy nature, extra time for transmission will be imposed, at the discretion of the umpire[32].

Telegraphy Cables- In order that a cable may be considered as cut, a telegraphy ship must have been on the spot and able to remain there for:

> 4 hours if under 30 fathoms
>
> 8 hours if between 50-100 fathoms
>
> 10 hours if between 100-300 fathoms
>
> 18 hours if over 300 fathoms.

For ordinary ships fitted with grapnels, add 50% to the time, provided the depth is less than 200 fathoms: over that depth only a properly fitted ship can cut the cable.

---

[31] I have added a map over the page to give an idea of the scale and strategic importance of this cable network to the British Empire.

[32] Actually, transmission times by cable were very rapid, see the Porthcurno Telegraphy Museum, details at www.porthcurno.org.uk/ JC.

Appendix IV Destroyers

Speed Allowances for Destroyers

Flotilla Leaders	Can steam in knots	Duration in hours
Scott Class	32	25
Parker Class	29½	25
Destroyers		
V and W Class	31½	27
R and S Class	32	22
Yarrow T.B.D.s	34	17
M Class	30	20

After which destroyers require to lay up for 24 hours in each case. Also, with a flotilla of 20 vessels, probably 4 would always be laid up for 4-5 day boiler-cleaning periods.

Flotilla Leaders	Maximum See Speed
Scott Class	33 ½
Parker Class	31½
Destroyers	
V and W Class	33½
R and S Class	34
Yarrow T.B.D.s	36
M Class	32

Speed in Rough Waters The speeds given above must be reduced considerably in these circumstances.

Reports from the Grand Fleet in 1918 placed the various classes in the following order of seaworthiness in heavy weather.

> S Class
>
> V Class
>
> R Class
>
> V Class (Leaders)

M Class

S Class kept up 20-24 knots with a sea force 4 on the bow, and R Class made 20 knots dead to windward without sustaining damage.

Increasing Speed

Speed may be increased at the following rates:

Speed	Knots per Minute
10-16 knots	3
16-20	2
20-24	1
24-28	½
28 upwards	¼

Putting Out of Action As in tables A and M.

If destroyers come under the fire of heavy guns inside 18,000 yards one destroyer will be considered to be put out of action every four minutes by each ship firing, if that ship is herself unfired at.

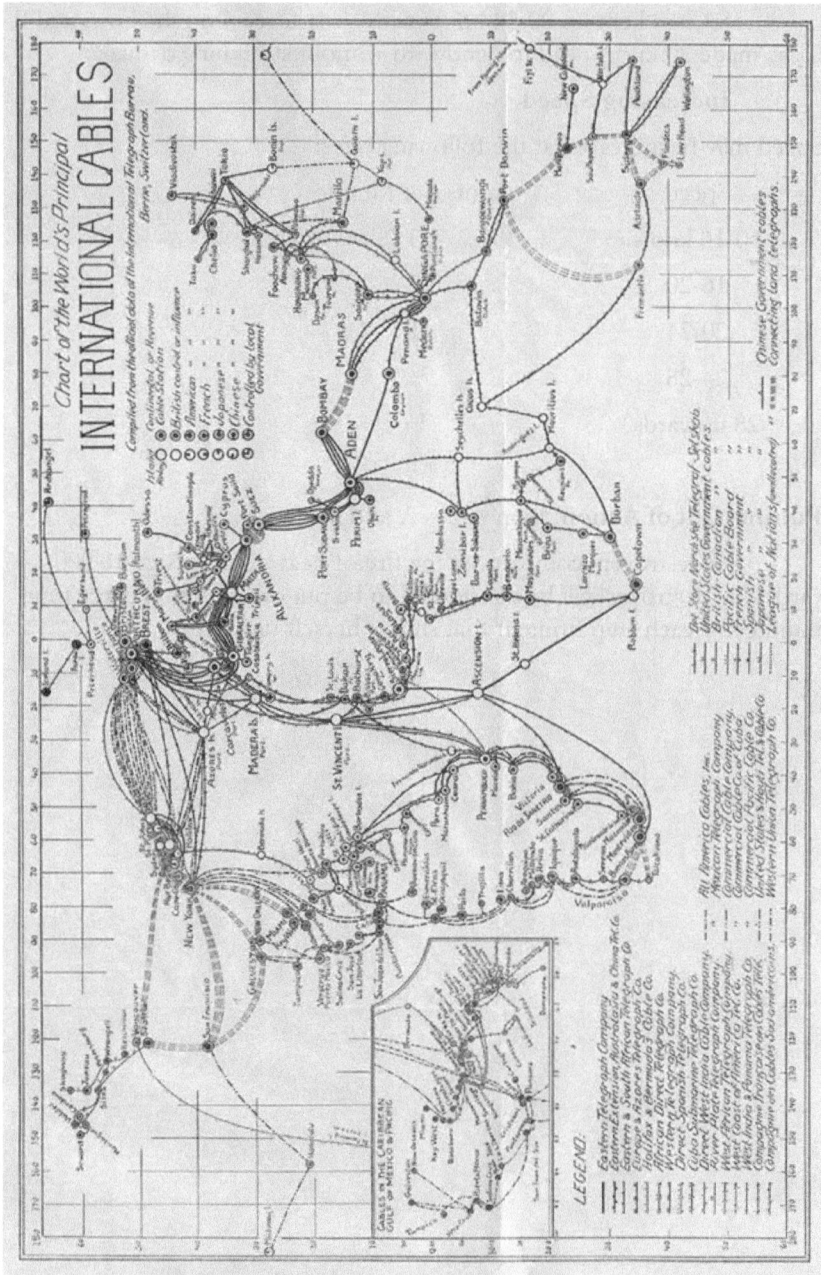

The British Underwater Cable Network

Vast, complex and fast. The results of the boat match between Oxford and Cambridge would reach India within a few hours. JC

Appendix V Submarines

Armament

Class	Guns	Torpedo Tubes	Torpedoes Carried
K	1-4 in. 1-3 in. H.A.[33] or 2-4 in.	8-18 in.	16
M	1-12 in. 1-3 in. H.A.	4-18 in. (M.3. 4-21 in.)	8
J	1-4 in. 1-3 in. H.A.	6-18 in.	12
L. 50	2-4 in. 1-3 in. H.A.	6-21 in.	12
L	1-4 in. 1-3 in. H.A.	Ls 1-8 4 -18 in. Later Ls 4-21 in. 2-18 in.	8 8-21 in. 4-21 in.
H	1-4 in. 1-3 in. H.A.	4-21 in.	6
R	1-4 in. 1-3 in. H.A.	6-21 in.	8

All classes carry one or two Lewis guns.

---

[33] H.A. = high altitude i.e. and AA Gun. JC

Above-Water Speeds, &c.

Type	Fuel Stowage in Tons. Normal Max.	Full Speed Knots	Patrol Endurance at Max. Stowage		Passage Endurance at Max. Stowage	
			Speed	Miles	Speed	Miles
H.21	14/16	11½	11½	1,520	11½	1,780
			9	2,310	9	3,420
R	17/19	9	9	1,660	9	2,200
			8	1,865	8	2,500
L.1 (Non-minelayer)	?³⁴4/76	17	17	2,542	17	2,760
			13	3,722	13	4,361
			10	3,890	10	4,865
L.1 (Minelayer)	??/??	13	13	2,270	13	2,460
			12	3,470	12	4,076
			10	3,930	10	4,920
L.50	78/100	15	15	2,950	15	3,190
			12	4,530	12	5,290
			10	5,220	10	8,100
J	7?/150	18½	18½	3,740	18½	3,950
			12	7,680	12	9,220
M.1	??/76	14½	Not used		14½	2,360
					10	1,320
M.2 and J	7?/114	14½	Not used		14½	3,340
					10	8,150
K	197/300	23	Not used		23	1,840
					10	4,000

---

[34] Some of the smallest characters failed to come out my copy of the original, ? means I could not read the missing digit. JC. If anyone can supply the missing information, I will add it on the website www.johncurryevents.co.uk

Under-Water Speeds

Endurance in hours

Class	H	R	L	J	K	M
Full Speed	9	14	10	9	9	9
14		1.3				
13		1.7				
12		2.2				
11		3				
10	1.0	4.2	2			
9	1.7	5.5	3	2	0.9	2
8	2.2	8.5	5	3.5	1.0	3.5
7	4	16	6.5	7	1.7	7.5
6	6.5	23	10	12	2.7	14
5	15	35	17	23	4.2	23
4		45	25	33	7.5	33
3	40			48	15.7	48
2.5						

Towing H and R Class submarines may be towed in smooth water.

Fuel Fuel can be obtained from any source, either ship or shore, where oil fuel is used.

Diving A submarine under way using her main engines on the surface will take 1½ minutes to dive, with the exception of K Class submarines which take 5 minutes to dive.

If a submarine is trimmed down she will dive in 30 seconds.

At night a submarine is usually on the surface and charging[35].

K Class submarines should be considered as surface craft if surprised on the surface.

Visibility The distance that a submarine can sight a ship, or a ship the submarine, will be settled by the umpire, who will then decide each case upon its merits.

Method of Moving and Assessing Results of Attack The track or position of submarines will be sent in by the officer in charge of the submarines when called for.

When the course of an enemy is such that she will be sighted by a submarine the Chief Umpire, will decide whether a successful attack is possible, taking into consideration the under-water speed of the submarine and the course and speed of the ship.

If successful attack is possible, the Chief Umpire will decide by drawing lots whether the ship sees the submarine or not, whilst the latter is delivering her attack.[36]

If the ship does not see the submarine, lots will be drawn to decide whether the torpedoes hits, the same chances being given as for a ship's short-range torpedo. (See Appendix II)

If the ship sees the submarine, the Officer in Command of the ship is informed of the position of the submarine. If he alters course, the action of any other submarine in the vicinity will be considered by the Chief Umpire in the same way.

---

[35] i.e. charging her batteries. JC

[36] The chances given of a ship seeing a submarine or her periscope before the torpedoes are fired will depend on weather conditions, and other conditions prevailing at the time.

Breaking Down The chances of a submarine breaking down and being unable to dive, or being unable to remain under water, will be left to the discretion of the umpire, who will decide each case upon its merits.

Rules for placing Submarines out of Action

(a) Gunfire A submarine in surface trim will be placed out of action if within 2,000 yards or any vessel for one minute; or within 4,000 yards of a ship or 3,000 yards of torpedo craft for two minutes.

(b) Depth Charge Attack A submarine actually sighted by a properly constituted hunting flotilla at a less distance than 2,000 yards has an even chance of escape.

At greater sighting distances

Sighting Distance	Chance of escape
4,000 yards	3 out of 4
6,000 yards	7 out of 8

A submarine sighted by escorting or patrol vessels, carrying depth charges, at less distance than 2,000 yards has 5 chances out of 6 to escape.

Appendix VI. Mining and Mine-Sweeping

Laying Mines

INFORMATION REQUIRED Tracings of all minefields are to be handed in to the Chief Umpire when called for.[37]

These tracings are to show the positions of the lines of mines and should be taken off the largest scale chart available.

The following information is also required whenever minefields are laid:-

> Distance apart of mines,
>
> Depth below IWOS,
>
> Number and type of mines,
>
> Type of sinkers used,
>
> Whether fitted with sinking plugs or delayed release,
>
> And the periods set for,
>
> By whom laid,
>
> Time taken in laying,
>
> Estimated time and date of completion.

RECORD OF NUMBER OF MINES   A record of the total number of mines laid is to be kept.

---

[37] For further information see Vol. 1, Part II, Mining Manual, 1920.

Table of Types of Mines, their Application, Main Features and Tactical Particulars (Part 1)				
Type of mine and sinker	HII Mine Mk II Mk VIII VIII*, VIIII*	II Mk Mk II Mk XII, XII*	H IIA Mine, Mk VIII VJTI*, VIII	S. IV Mine on sinker
Charge in Mine (nominal)	Amatol 320	Amatol 320	Amatol 320	Amatol 210
Mooring wire	60 fms 1¼ Mk VIII. 175 fms 9¼ Mk VIII* . 215 fms 1, Mk VIII*	60 fms 1½	60 fms 1½ Mk VIII. 175 fms, .94 Mk VIII*. 216 fms, 1 MK VIII*	30 fms 1¼
Special Devices	Sinking plug	Sinking plug Delayed release	Sinking plug	Sinking plug
Active	At all times when moored	At all times when moored	At all times when moored*	At all times when moored
Dispersal when laid	150 feet	150 feet	150 feet	120 feet

*The 'A' Attachment looses effectiveness if moored below 10 fathoms and fails altogether at 30 fathoms; also it is normally safe if moored at less depth below surface than 25 feet. The contact firing gear remains effective in all these cases.

Table of Types of Mines, their Application, Main Features and Tactical Particulars (Part 2)			
Type of mine and sinker	S. V Mine on sinker	S. VA Mine on sinker	L Mine on Sinker
Charge in Mine (nominal)	Amatol 250	Amatol 210	Amatol 500
Mooring wire	30 fms 1¼	30 fms 1¼	40 fms paravane cable 11/16
Special Devices	Sinking plug		Can take 'A' attachment only when current supplied externally
Active	At all times when moored	At all times when moored	
Dispersal when laid	150 feet	150 feet	150 feet

*The 'A' Attachment looses effectiveness if moored below 10 fathoms and fails altogether at 30 fathoms; also it is normally safe if moored at less depth below surface than 25 feet. The contact firing gear remains effective in all these cases.

Table of Types of Mines, their Application, Main Features and Tactical Particulars (Part 3)

Type of mine and sinker	HII Mine Mk II Mk VIII VIII*, VIIII*	II Mk Mk II Mk XII, XII*	H IIA Mine, Mk VIII VJTI*, VIII	S. IV Mine on sinker
Charge in Mine (nominal)	Amatol 320	Amatol 320	Amatol 320	Amatol 210
Means of obtaining depth	Placement	XII fixed moorings XII hydro static	Placement	Hydrostatic
Operation for which designed	Automatic laying from AW mine layers	Drop laying and automatic laying with hydrostatic AW mine layers	Automatic laying from AW mine layers	Automatic laying from E Class submarines and 70 ft CMH's
Max. Depth mine will stand	70 fms	70 fms	70 fms	40 fms
Min. Depth which mine can be laid- The mine being 8 ft below surface at low water	5 fms	3 fms	5 fms	3 fms

Note: submarines cannot lay in 3 fms.

Table of Types of Mines, their Application, Main Features and Tactical Particulars (Part 4)			
Type of mine and sinker	S. V Mine on sinker	S. VA Mine on sinker	L Mine on Sinker
Charge in Mine (nominal)	Amatol 250	Amatol 210	Amatol 500
Means of obtaining depth	Hydrostatic	Hydrostatic	Fixed moorings
Operation for which designed	Automatic laying from L Class submarines and 70 ft CMH's	Automatic laying from L Class submarines and 70 ft CMH's	Controlled mining L system from special L minelayers
Max. Depth mine will stand	70 fms	70 fms	75 fms
Min. Depth which mine can be laid- The mine being 8 ft below surface at low water	3 fms	3 fms	Blank

POINTS TO BE OBSERVED – The following points should be observed.

(i) A minefield must start at a safe distance from those previously laid in the vicinity.[38]

(ii) Minelayers must return to harbour to embark another outfit of mines.

(iii) Speed of mine laying. The actual minimum working interval (in time) between successive mines laid from the same rail is 12 seconds.

---

[38] For further information, see Volume L. Part II, Mining Manual, 1920.

For A.W. minelayers other than CMB's a speed of 15 knots should not generally be exceeded on account of the stern wave introducing errors in depth taking; for the same reason, the status of the sea must be taken into account, errors of depth equal to half the height of the wave may be expected. A table of time intervals for various speeds is given in Torpedo Manual Vol. 1 Part II.

Submarines are limited to a speed of 5 knots while laying to ensure the mines clearing the hull on release.

CMB's usually lay in groups and stop engines before laying.

(iv) Mines are inoperative till one hour after laying.

(v) The minefield efficiency of all contacts mining systems is affected:-

(a) Nearly always by *rise* and *fall* of tide and

(b) Frequently by the strength of the current.

In order to assess Minefield Efficiency, it is necessary first to prepare a Minefield Efficiency Diagram and it is always advisable that this should be done for all minefields unless the laying conditions are similar.

(vi) Minelayers cannot lay mines under the following weather conditions:-

A wind of force 7 or more from any direction, or a sea of force 5 or more on the quarter. These figures may be reduced by 25% when destroyer minelayers are employed.

(vii) When passing through a line of mines, the chances of a ship striking a mine will be decided by the Chief Umpire.

(viii) Number of mines carried.

a) Large A.W. Minelayers according to class of vessel.

b) Destroyer minelayers , 40-86 H. II mines according to class of vessel.

c) (1) E Class Submarines 20 S IV. mines

(2) L Class submarines 16 S V. mines

d) Monitor minelayers, 48 H II. Mines and L mines if specially fitted.

e) 40 ft CMB.'s 1 S. IV mines.

55 ft C.M.B.'s 4 S. V. mines.

70 ft C.M.B.'s V. mines.

(ix) Mines should be considered active at all times when moored.

The rated times for sinking plugs are:

8 hour

60 hour

10 day

38 day

Mine Sweeping

Fleets must depend on paravanes as a protection from mines laid in the open sea.

There are no such ships as fleet minesweepers in the true sense i.e. vessels capable of accompanying the fleet to sea sweeping ahead. Destroyers using H.S.M.S. come nearest to this definition, but the sweep is not available.

The functions of fleet sweepers are:

(i)   to search and keep clear the channels giving exit and entrance to the fleet bases;

(ii)   to define and/or clear enemy minefields;

(iii)   to skim your own deep minefields.

Sweeping of exits and entrance channels takes place daily or periodically or immediately before use.

The sweeping of the channel to be used must be taken into consideration when fixing he time of sailing of the fleet.

Minefields found outside the fleet channels must be defined by the fleet sweepers and their limits reported, for the Commander-in-Chief to decide whether it should be cleared at once or later, or left down as a defensive minefield.

While it is necessary to route ships over your own deep minefields it will be necessary to skim these areas to remove enemy shallow mines laid subsequently.

Fleet sweeping therefore, takes place independently of the fleet, except in particular operations as:

 (i) Where the speed of the fleet can be safely reduced to admit of them keeping station astern of the sweepers.
 (ii) Bombarding operations by monitors at slow speed.
 (iii) Some minelaying operations.
 (iv) In special cases in shallow water ahead of a convoy.

It may be assumed that no mines can be laid in water of greater depth than 200 fathoms.
It may be assumed that nine twin-screw minesweepers of the Aberdare class are attached.
(i) To each fleet.
(ii) to each fleet base.
   And 12 trawlers (a) to each fleet base
        (b) to each auxillary base.

Two thirds of the total strength are always ready for sea.

TWIN-SCREW MINESWEEPERS.
  Maximum sweeping speed, 9 knots.
  Maximum station keeping speed running free, 14 knots.
  Maximum draft, 8 ft.
  Breadth of sweep of one pair , 2½ cables.
  One pair can sweep 2.25 square miles per hour.

TRAWLERS

  Maximum sweeping speed, 6 knots.
  Maximum station keeping speed running free, 9 knots.
  Maximum draft, 15 ft.
  Breadth of sweep of one pair , 2 cables.

One pair can sweep 1½ square miles per hour.

SEARCHING SWEEPS

A percentage of the water only is swept. 50% is considered sufficiently high percentage to ensure discovery of a minefield laid in a channel.

The chances of the discovery of scattered or isolated mines vary directly as the percentage of water swept.

CLEARING SWEEPS

(A) By T.S.M.S.[39]:

(i) By three pairs in D formation in water of less depth than 20 fathoms. Area cleared = 4 square miles per hour.

(ii) In water of more than 20 fathoms sweeping in F formation i.e. line abreast. Area cleared per hour is 1.5 * (x-2) square miles where x = number of minesweepers employed.

b) By Trawlers:

(i) By three pairs in D formation in water of less depth than 20 fathoms. Area cleared = 2 square miles per hour.

(ii) In water of more than 20 fathoms sweeping in F formation i.e. line abreast. Area cleared per hour is 0.9 * (x-2) square miles where x = number of trawlers employed.

DELAY RELEASE

This can be countered by trawlers using the bottom sweep at speed of 4 knots.

CASUALTIES

One T.S.M.S. or two trawlers for every 50 mines swept up.

These casualties need not be incurred if sweeping can be confined to safe hours when:

---

[39] Twin Screw Mine Sweepers. JC,

a) The rise of the sea level above L.W.O.S. is greater than the draught of the sweepers.

b) The dip of the mine caused by the tidal stream is greater than the draught of the sweepers.

c) A combination of (a) and (b).

WEATHER

Searching sweeps up to sea state 6[40].

Clearing sweeps up to sea state 4[41].

For further details and information, see 'Handbook of Minesweeping'.

---

[40] Sea State 6 on the Beaufort scale is 'Large waves begin to form; the white foam crests are more extensive everywhere. Probably some spray'.

[41] Sea State 4 on the Beaufort scale is 'Small waves, becoming larger; fairly frequent white horses'.

Appendix VII: Aircraft

Aircraft may be used under the following conditions:

INFORMATION REQUIRED: a tracing, showing their tracks and successive positions as described in the Rules for Strategical Exercises is to be sent in with the following modifications.

The course steered and the drift due to the wind allowed for are to be shown in red and the course made good in black.

The speed of the aircraft and the speed of the wind are to be noted against the course steered and the line showing the drift.

Positions where wind was verified are to be shown in the case of airships.

On receipts by the Chief Umpire, the actual position reached by the aircraft, allowing for all errors, will be marked on the Umpire's chart and this position, as far as it could be known to the aircraft, will be marked on the tracing when the latter is returned in readiness for the next move.

BREAKDOWN- Every 24 hours the Chief Umpire will draw lots for the chances of aircraft breaking down. The chances given will depend on weather, class of aircraft, and other conditions prevailing at the time.

Airships

SPEED – Speed through still air 45 knots maximum for 48 hours, at the end of which period they must be at a base and will be unable to proceed for eight hours and until they have replenished with fuel and gas. When there is wind, the speed of the airship on any course is the resultant of the speed of the wind and speed of airship.

BASES - She may operate from three descriptions of base:

(a) Permanent base. Safe under all conditions.

(b) Temporary base. Requires at least 1 fathom of water and clearing swinging room and takes 18 hours to prepare with the resources of a dockyard, large seaport or battle division. Safe in a wind not exceeding force 6.

ERRORS IN POSITION. An airship can ascertain the direction and speed of wind by stopping for a quarter of an hour in every four hours. If this is not done, she will be liable to an error of two points and 15% of speed in her estimation of the wind when out of sight of land. She will also be liable to a further error of 10 miles in every two hours travel when out of sight of land.

WIRELESS TELEGRAPHY: Range for wireless telegraphy, 500 miles.

Aeroplanes and Seaplanes

SPEED Details of speed in still air are given in the attached table; if there is wind, the speed of the aeroplane on any course is the resultant of the speed of the wind and speed of that aeroplane.

At the end of the period of endurance given, an aeroplane must be back on land, or on a properly equipped carrier e.g. Eagle or Argus and cannot proceed again for half an hour.

ERRORS IN POSITION An aeroplane is liable to an error in position of 5 miles in every 2 hours travel when in sight of land, or 10 miles in every 2 hours travel when out of sight of land.

WIRELESS TELEGRAPHY Range (if fitted), 150 miles.

Existing Types of Aeroplanes and Seaplanes

Type	No. of Seats	Max. Speed knots	Endurance at cruising speed (hours)	Armament	Remarks
Aeroplanes					
Panther	2	95	4½	2 machine guns	Reconnaissance and observation of fire. Can fly off and land on a carrier or fly of fore-turret
Night Hawk	1	122	2½	2 machine guns	Fighter and attacking exposed personnel. No signalling gear, can fly off and land on a carrier or fly of fore-turret
Cuckoo	1	87	3½	1 18" torpedo	Torpedo carrier. No signalling gear, can fly off and land on a carrier.
Boat Seaplanes					
F.II.A.	5	83	7½	On	Reconnaissance

F.V.	5	89	8½	submarine patrol, 2 machine guns, about 1,000lb of bombs	and bombing

Float Seaplanes

Short	2	73	4	-	Reconnaissance, bombing and observation of fire. Can only rise in very sheltered water. Poor climbers and very vulnerable to attack by fighters.
Fairey	2	96	5½	1 machine gun about 300 lb weight of bombs	

Aircraft Carriers

Name	Guns	Torpedo Tubes	Speed	Date of completion
Argus	2 4" 4 4" H.A.	-	20-21	1918
Eagle	12 6" 4 4" H,A.	18 A.W. 21"	24	1920
Hermes	10 5.5" 4 4" H.A.	-	26	Building

Appendix VIII: Major Warship Details[42]

Details of ships completed not earlier than 1910

N.B. Also details of latest Flotilla Leaders and Destroyers (minelayers and auxiliary vessels are not included). The bearings of foreign torpedo tubes and guns, when given, are only approximate. Tubes are 18 inch (approximate) unless otherwise stated.

Ship	Date of completion	Full speed	Guns	Arc of fire from 0⁰ ahead to 180⁰ astern	Submerged tubes and their bearings (above water tubes marked A.W.	No. torpedoes carried
Great Britain Battleships						
5 Royal Sovereign (Royal Oak, Ramillies, Revenge, Resolution)	1916-1917	22	8 15" L 14 6" XII	C 30 A 150	4 beam 21"	20
5 Queen Elizabeth (Malaya, Warspite, Barham, Valiant)	1915-1916	25	8 15" I 12 6" XII	C 25 B 30 A 150	2 10⁰ B 2 10⁰ A both 21"	20
4 Iron Duke (Emperor of India, Benbow, Malborough)	1914	21	10 13.5" 12 6" VII	C 40 A 140 B 150	2 10⁰ B 2 10⁰ A both 21"	20
1 Erin	1914	21	10 13.5" 16 6"	C 30 A 140 B 150	2 5⁰ B 2 5⁰ A both 21"	16
3 King George V (Ajax, Centurion)	1912-13	21	10 13.5" V	C 40 A 140 B 150	2 10⁰ B 21"	14

---

[42] Personally, I would refer to Jane's Fighting Ships for details and not rely on this appendix. JC.

Ship	Date of completion	Full speed	Guns	Arc of fire from 0⁰ ahead to 180⁰ astern	Submerged tubes and their bearings (above water tubes marked A.W.	No. torpedoes carried
4 Orion (Conqueror, Monarch, Thunderer)	1912	21	10 13.5" V.	C 40 A 140 B 150	2 beam 21"	14
2 Colosssus (Hercules)	1911	21	10 12" X.I.	D 40 Starboa	2 beam 21"	14
1 Neptune	1911	21	10 12" XI.	rd B 50 A 110 B Port B 70 A 130 B 140 C	2 beam	12
2 St. Vincent (Collingwood)	1909-10	21	10 12" XI	C 40 A 140 B165 D	2 15⁰ B	9
Great Britain Battle Cruisers						
1 Hood	1920	31	8 15" L 12 5.5" I.	C 30 A 150 C	2 Beam 4 A.W. all 21"	20
2 Renown (Repulse)	1916	31.5	6 15" I		2 beam 21"	10
1 Tiger	1914	28	8 13.5" V.12 6" VIII	C 30 A 150 C	2 10⁰ beam 2 10⁰ aft all 21"	20
2 Lion (Princess Royal)	1912	27	8 13.5" V	C 40 A 140 B 150 C	2 beam 21"	14
1 New Zealand	1912	25	8 12" XI	C 30 St B 50 A 110 B Port B 60, A. 130 B.	2 beam 21"	12

Ship	Date of completion	Full speed	Guns	Arc of fire from 0⁰ ahead to 180⁰ astern	Submerged tubes and their bearings (above water tubes marked A.W.	No. torpedoes carried
				140 C		
Great Britain Cruisers						
2 Courageous (Glorious)	1917	31	4 15" I	C 30, A 150 C	2 beam 12 A.W. all 21"	22
1 Furious	1917	31	10 5.5" I	-	2 beam 16 A.W. all 21"	25
Great Britain Light Cruisers						
4 Hawkins (Raleigh, Effingham, Frubisher)	1919 and later	30	7 7.5 I	C 30, B 48, A 135, B. 150 C	2 beam 4 A.W. all 21"	12
1 Vindictive	1918	30	4 7.5 I	-		15
2 Emerald (Enterprise)	Building	33	7 6" XII	-		-
8 Dragon (Delhi, Dunedin, Danse, Dauntless, Durban, Despatch, Diomede	1918 and later	29	6 6" XII	C 20, B 35, A 145, B 155 C	12 A.W. 21"	13
5 Carlisle (Cairo, Calcutta, Columbo, Capetown)	1918 and later	29	5 6" I	C 20, B 35, A 145, B 160 C	8 A.W 21"	8
5 Ceres (Cardiff, Coventry, Curacoa, Curlew)	1917-18	29	5 6" XII	-	8 A.W 21"	8 or 9

Ship	Date of completion	Full speed	Guns	Arc of fire from 0⁰ ahead to 180⁰ astern	Submerged tubes and their bearings (above water tubes marked A.W.	No. torpedoes carried
5 Caledon (Calypso, Caradoc)	1917	29	5 6" XII	C 20, B 50, A 130, B 145, C.	8 A.W 21"	8
2 Centaur (Concord)	1916	28.5	5 6" XII	-	2 beam 21"	7
6 Cambrian (Canterbury, Castor, Champion, Calliope)	1915-16	28.5	4 6" I	-	2 beam 21"	7
Caroline (Cleopatra, Carysfort, Cordelia, Comus, Conquest,	1914-15	28.5	4 6" XII	-	4 A.W. 21"	4
7 Arethusa (Royalist, Inconstant, Penelope, Galatea, Undaunted, Phaeton)	1914-15	28.5	2 or 3 6" XII	-	8 A.W. 21"	9
2 Birmingham (Lowesestoft)	1912-14	25.5	9 6" XII	-	2 beam 21"	7
3 Chatham (Dublin, Southampton)	1912-13	25.5	8 6" XI	-	2 beam 21"	7
3 Dartmouth (Weymouth, Yarmouth)	1911-12	25	8 6" XII	-	2 beam 21"	7
Great Britain Flotilla Leaders and Destroyers						
Scott Class	1918-19	36.5	5 4.7" I	-	6 A.W. 21"	6

Ship	Date of completion	Full speed	Guns	Arc of fire from 0^0 ahead to 180^0 astern	Submerged tubes and their bearings (above water tubes marked A.W.	No. torpedo es carried
Parker Class	1916	34	4 4" IV	-	4 A.W. 21"	4
V Claa	1917	34	4 4" V	-	6 A.W. 21"	6
Mod. W Class	1919	34	4 4.7 I	-	6 A.W. 21"	6
W Class	1917-18	34	4 4" V	-	6 A.W. 21"	6
V Class	1917-18	34	4 4" V		4-6 A.W. 21"	4-6
S Class	1918-19	36	3 4" IV	-	4 A.W. 21"	4
R Class	1916-17	36	3 4" IV	-	4 A.W. 21"	4
M Class	1914-16	34	3 4" IV	-	4 A.W. 21"	4
Australia Battle Cruisers						
1 Australia	1913	25	8 12" XI	C 40 St B 50 A. 110 B. Pt B 60, A 130 B 140 C	2 beam	18
Australia Light Cruisers						
1 Adelaido	Building	25.5	9 6" XII	-	2 beam 21"	8
3 Brisbane (Sydney, Melbourne)	1913	25.5	8 6" XI	-	2 beam 21"	8
France Battleships						
3 Bretagne (Province, Lorraine)	1915-16	20	10 13.4" 22 5.5"	C 30, B 45, A 135, B 150 C	4 beam	12
4 Jean Bart (France, Paris, Courbert)	1913-14	20	12 12" 22 5.5"	C 40, B 45, A 135,	4 beam	12

Ship	Date of completion	Full speed	Guns	Arc of fire from $0°$ ahead to $180°$ astern	Submerged tubes and their bearings (above water tubes marked A.W.	No. torpedoes carried
				B 140 C		
4 Condorcet [43](Vergniaud, Voltaire, Diderot)	1911	19	4 12" 12 9.4"	C 15, B 45, A 135, B 180 C	2 beam	4
French Cruisers						
2 Falgar (Waldock, Remsean?)	1911	23	14 7.6"	C 10, B 60, A 120, B 170 C	2 beam	-
France Light Cruisers						
1 Strasburg	1914	27.5	7 5.9"	-	2 A.W. 19.7"	-
1 Metz	1916	28.5	8 5.9"	-	2 beam 2 A.W. all 19.7	-
1 Mull?	1912	27	7 5.9"	-	2 beam 2 A.W. all 19.7	-
1 Colmar	1910	25	6 5.9"	-	2 A.W.	-
1 Thionville	1914	27	8 3.9"	-	8 A.W.	-
Ex German	1917-18	34	3 4.1"	-	6 A.W.	-
Koba Class	1917	29	14.7" 4 12 pdrs	-	4 A.W.	8
Temeraire Class	1914	32	4 3.9"	-	4 A.W.	-

---

[43] Became Danton Class. JC.

Ship	Date of completion	Full speed	Guns	Arc of fire from 0^0 ahead to 180^0 astern	Submerged tubes and their bearings (above water tubes marked A.W.	No. torpedoes carried
Commadant Bory Class	1913-15	31	2 3.9" 4 9pdr	-	4 A.W.	6
Italy Battleships						
2 Andreas Doria (Duilio)	1915-16	22	13 12" 16 6"	C 25, B40 A 140, B155 C	2 40⁰ B 1 Stern	-
2 Giulio Cesare (Conte di Cavour	1914-15	22	13 12" 18 4.7"	C25, B30, A 150, B 155 C	2 40⁰ B 1 Stern	-
1 Danto Aligherit	1912	22	12 12" 20 4.7"	C 25, B 35, A 145, B155 C	2 beam ? 1 stern	-
Italy Cruisers						
3 San Marco (San Giorgio, Pisa)	1909-10	23	4 10" 8 7.5"	C 20, B 35, A 145 C (Pisa 55 A. 130)	2 beam ? 1 stern	-
Italy Light Cruisers						
2 Basilicata (Campania)	1916-17	16	6 6"	-	2 A.W.	-
3 Nino Bixio (Quarto, Marsala)	1912-14	27	6 4.7"	-	2 A.W.	6
1 Labia	1913	22	2 6" 8 4.7"	-	2 A.W.	-

Ship	Date of completion	Full speed	Guns	Arc of fire from 0^0 ahead to 180^0 astern	Submerged tubes and their bearings (above water tubes marked A.W.	No. torpedoes carried
1 Ancona	1914	27.5	7 5.9"	-	2 A.W. 19.7"	-
2 Brindisi (Venezia)	1914	27	8 3.9"	-	2 A.W.	-
1 Bari	1914	27.5	8 5.9"	-	6 A.W. 19.7"	-
1 Taranto	1912	27	7 5.9"	-	2 beam 2 A.W. all 19.7"	-
Italy Flotilla Leaders and Destroyers						
Aquilia Class	1916-19	35	5 4.7" 4 14pdr		4 A.W.	-
Mirubello Class	1915-16	35	8 4"		4 A.W.	-
Sirtori Class	1916-18	33	8 or 4 4"		4 A.W.	-
Abba Class	1915-16	30	5 4"		4 A.W.	-
Japan Battleships						
2 Kaga (Tosa)	Building	23	8 or 10 16" 20 5.5"	-	8 ?	-
2 Nagato (Matsu)	Building	23	8 16" 22 5.5"	-	8 ?	-
2 Hyuga (Iso)	1917-18	23	12 14" 20 5.5"	C 30, A 140, B 150 C	6 21"	12
2 Yamashiro (Fuso)	1915-17	22.5	12 14" 20 5.5"	C 30, A 45, B 150, B 155 C	6 21"	12
1 Settsu	1912	20.5	12 12"	C 20,	4 beam	-

Ship	Date of completion	Full speed	Guns	Arc of fire from 0⁰ ahead to 180⁰ astern	Submerged tubes and their bearings (above water tubes marked A.W.	No. torpedoes carried
			10 6" 8 4.7"	B 35 A 145, B 160 C	1 stern	
1 Aki	1910	20	4 12" 12 10" 8 6"	C 30, B 42, A 138 B 150 C	4 beam 1 stern	15
1 Satsuma	1910	18	4 12" 12 10" 12 4.72	C 30, B 42, A 138 B 150 C	4 beam 1 stern	15
Japan Battle Cruisers						
4 Akago (Amagi, Takao, Atago)	Projected	-	-	-	-	-
4 Kongo (Hiyoi, Haruna, Kirishima)	1913-15	27	8 14" 16 6"	C 20, B25, A 150 C	8 beam 21"	24
2 Kurama (Ibuki)	1909-10	22	4 12" 8 8" 14 4.72	C 30, B 40, A140 B 150 C	2 Beam, 1 stern	9
Japan Light Cruisers						
4 unknown	Projected	-	-	-	-	-
3 Kitakami (Kiso, Oi)	Building	-	-	-	-	-
2 Kam ??	Building	33	7 5.5"	-	8? A.W. 21"	-

Ship	Date of completion	Full speed	Guns	Arc of fire from 0^0 ahead to 180^0 astern	Submerged tubes and their bearings (above water tubes marked A.W.	No. torpedoes carried
(Tam??)						
2 Tabanta (Tenryu)	1919	31	4 5.5"	-	6 A.W. 21"	-
3 Yahagi (Hirado, Chikuma)	1912	26	8 8"	C 30, B 45, A 135, B 150 C	3 A.W.	-
Japan Destroyers 1st Class						
Hakaze Class	Building	34	4 4.7"	-	6 A.W. 21"	12 ?
Kawakazo Class	1918-19	34	3 4.7"	-	6 A.W 21"	12 ?
Ama?kaxo	1917	34	4 4.7"	-	6 A.W	12 ?
Umikaze Class	1910-11	31.5	2 4.7" 5 12pdrs	-	4 A.W	8 ?
Japanese Destroyers 2nd Class						
Moni Class	1919-20	31.5	3 4.7"	-	4 A.W 21"	8 ?
Momo Class	1917-18	31.5	3 4.7"	-	6 A.W	12 ?
United States Battleships						
6 South Dakota (Indiana, Iowa, Montana, Massachusetts, North Carolina)	Building	23	12 16" 16 6"	-	2 21"	-
4 Colorado (Maryland, Washington, West Virginia)	Building	21	8 16" 14 5"	-	2 21"	16 ?
2 Tennessee (California)	Building	21	12 14" 14 5"	-	2 21"	-
3 New	1917-19	21	12 14"	C 30,	2 21"	8 ? 12 ?

Ship	Date of completion	Full speed	Guns	Arc of fire from 0⁰ ahead to 180⁰ astern	Submerged tubes and their bearings (above water tubes marked A.W.	No. torpedoes carried
Mexico (Idado, Mississipi)			14 5"	B 35 A 145 C		
2 Pennsylvania (Arizona)	1916	21	12 14" 14 5"	C 30, B35, A 145, B150 C	2 21"	8 ? 12 ?
2 Nevada (Oklahoma)	1916	20.5	10 14" 12 5"	C 25, B 30, A 135, B 50, C	4 beam 21"	8
2 New York (Texas)	1914	21	10 142 16 5"		2 21"	12
2 Arkansas (Wyoming)	1912	20.5	12 12" 16 5"		2 21"	8
2 Utah (Florida)	1911	20.75	10 12" 16 5"		2 21"	-
2 Delaware (North Dakota)	1910	21	10 12" 14 5"		2 15" B. 21"	8
United States Battle Cruisers						
5 Lexington (Constellation, Saratoga, Ranger, Constitution, United States)	Building	33.25	8 162 16 6"		4 4 A.W. all 21"	
United States Light Cruisers						
10	Building	35	8 6"		4 A.W. 21"	-
Nos. 296 347	Building	35	4 4"		12 A.W. 21"	12
Nos. 231-235	Building	35	4 5"		12 A.W. 21"	12

Ship	Date of completion	Full speed	Guns	Arc of fire from 0^0 ahead to 180^0 astern	Submerged tubes and their bearings (above water tubes marked A.W.	No. torpedoes carried
Nos. 186 230 Nos. 236-295	1918-1920	35	4 4"		12 A.W. 21"	12
Nos. 75-185	1918-1919	35	4 4"		12 A.W. 21"	12

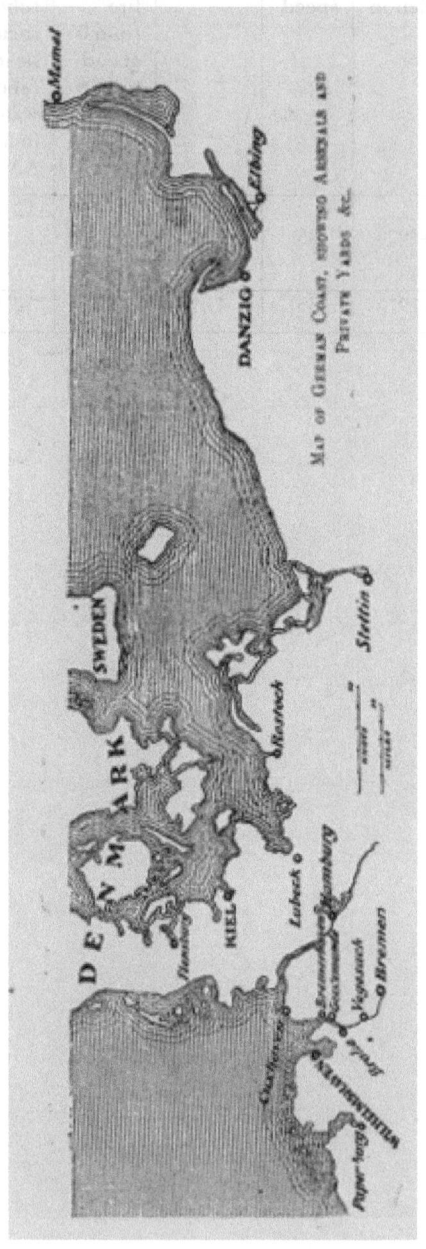

MAP OF GERMAN COAST, SHOWING ARSENALS AND PRIVATE YARDS &c.

Appendix: Your Navy as Fighting Machine by Fred T. Jane (1914)

YOUR NAVY AS A FIGHTING MACHINE

BY

Fred T. Jane

Founder Editor of
"Fighting Ships"

LONDON

FRANK & CECIL PALMER

12-14 RED LION COURT

October, 1914

CONTENTS

INTRODUCTION

This little book is an attempt to produce an entirely non-technical handbook for the use of those who, till European war came along, did not interest themselves in naval matters. Till now a vast number of people have taken the Navy for granted. It has existed to them much as St. Paul's Cathedral exists. To the great majority there has been no occasion to trouble about anything, save perhaps one or two of the more picturesque features of the Fleet. Now, however, after a hundred years of peace the British Navy is engaged in naval warfare and the entire situation is changed accordingly.

It is true that during this past century the Navy has been engaged in various operations. In the Crimean War, for instance, two considerable fleets were employed. Both before and since our ships have bombarded forts and places, like Algiers and Alexandria, but in all the hundred years there has been no war between British fleets and the fleets of a foreign Power. And so it comes about that all eyes are now upon the Navy, which somewhere on the seas started facing the unknown directly Austria sent her ultimatum to Serbia.

So soon as that incident occurred everything which has happened since became a vivid possibility. From that moment the Fleet had to be on watch and guard lest Germany should fall on us unawares. That she intended to attempt it was perfectly well-known it had been known for years to all in authority.

The British Navy for which the public has paid is now undergoing the supreme test.

FRED T. JANE.

I. THE CAPITAL SHIP.

What is a "capital ship"? It is a vague sort of term that is frequently employed; nor is much further information usually to be gleaned than that "it's a battleship."

If there be one thing more than another of which the general public is supremely ignorant that thing is a battleship. They may know its guns, they may know its displacement, speed and various other details the sort of thing to be found in every explanatory footnote in the daily Press. But it is safe to wager that not one man in a million in all the British Empire knows exactly what a battleship is for. All have a vague idea, because it so chances that people "think in Dreadnoughts" and they all know that Dreadnoughts are somewhere in the top line of "battleships."

Well, in actual fact the matter is somewhat like this. Super-Dreadnoughts, Dread-noughts, pre-Dreadnoughts, etc., etc., etc., are one and all precisely the same thing when it comes to the real point. At the best they merely represent certain theoretic fighting values.

Why they do so will be explained later. For the present it suffices to state that for fighting purposes the battleship is a battleship and that, despite the fact that a single super-Dreadnought could probably sink a dozen pre-Dreadnoughts there is still not the remotest difference between one kind of battleship and another, except in power once the business of war comes to the front.

The difference between one kind and another is on all fours with the difference between big and little soldiers.

Good, bad, or indifferent the duties of a battleship are and must ever be exactly the same thing just as they were half-an-hour ago or a hundred years ago or a thousand or more years ago.

The battleship has been all kinds of things in this interim. When our remote ancestors started the "battleship" idea they did so on tree-trunks. When Drake and Nelson (intervals between) did so there was no essential difference in main principles. There is no more at the present day. There will never be any real difference all through the aeons of time.

The battleship, whatever form she may take hereafter, will ever be the battleship, the equivalent of the Queen at chess, or of the Ace in cards.

Its specialty is to smash. Till a smashable stage is reached, it has nothing to smash and it does not smash accordingly. The value of an ace at cards is that no one plays it until the playing of it entails a distinct advantage not otherwise to be obtained. For example: no one would (if he could help it) play an ace to take a trick in which the highest previous card was a seven of something. Similarly, no chess player would risk his queen to effect a capture which a pawn could accomplish.

He would keep his ace to beat a king, or his chess queen to dispose of a rook.

War is just the same. That is why there was no attempt at a Trafalgar or a Tsushima in the early stages of the present war.

That is why the Dreadnought will only be used once. Just as an ace can only be played once, naval warfare can never be a matter of parading Dreadnoughts inside the danger zone.

Modern Trafalgars are not made that way. Nowadays as the pen is mightier than the sword, so the pen figuring things out mathematically is a mighty lot greater than anything else.

The battleship can only be used once used for the " knock-out blow." In a war in which both sides made no mistakes no fleet action could possibly occur, no battleship fire a shot in anger.

The highest use of a battle fleet costing easily fifty million pounds is, curiously enough, complete inaction! So long as unused it is worth all that fifty million. Once used its value may easily deteriorate to nothing or thereabouts. On one side or the other this is bound to occur. Whatever the value of an ace there is only one ace of trumps. Just the same thing occurs with battle fleets. Till the psychological moment arrives their only real use lies in disuse.

Out of which various folk who do not or cannot think much have described the "battleship" as "no use."

Were there no hostile battleships that, of course, would exactly describe the situation. Similarly, however, if people played without any Court cards in the pack an ace might easily be a superfluity !

To sum up, therefore, the precise duties of a battleship and the precise reasons why she exists are as follows:

(1) To smash others of her kind as the need arises.

(2) To be a rallying point on which weaker units can retire if over- pressed by others of their own kind. So long as the battleship can with certainty perform these functions it matters nothing whether she be Dreadnought or pre-Dreadnought, though naturally a Dreadnought is more likely to succeed.

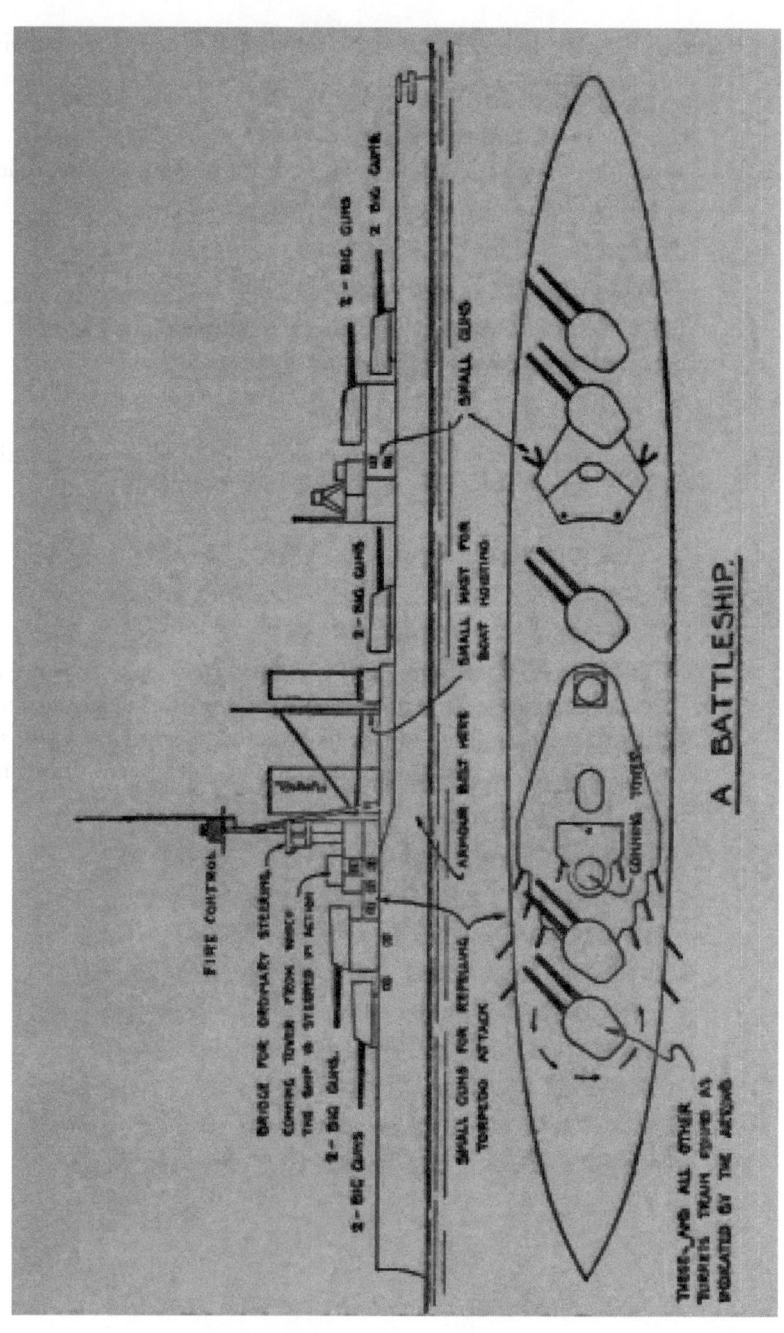

A BATTLESHIP.

WHAT IS A DREADNOUGHT?

Many years ago, back at the time of the American Civil War, there appeared a curious little ironclad with her guns in a turret, and lying very low in the water. She fought a famous battle in the Hampton Roads against a Confederate ironclad, the Merrimac, which was a ship more or less of the ordinary type, covered with iron plates. This curious little turret ship was named the Monitor, and thereafter all ships subsequently built which in any way resembled her were spoken of as Monitors.

In just the same way for the last few years every battleship which carries more than four big guns is christened a Dreadnought, because that is the name of the first "all-big-gun-ship" built for the British Navy.

Except that she is more powerful there is no integral difference between a Dreadnought and the older kind of battleships which are generally spoken of as pre-Dreadnoughts. Ten years ago all our battleships carried four big guns (i 2-inch), two in a turret forward, and two in a turret aft. The rest of the guns were 6-inch.

After a time, however, it was thought that something bigger than the 6-inch was required. The Germans mounted guns of 6.7 -inch, in what is technically known as the secondary battery secondary battery being a general term for the guns of less importance than the main armament.

Italy, the United States and one or two other nations hit on the idea of having some of their secondary guns 8-inch or thereabouts, while others were 6-inch.

In the British Navy, however, it was thought that these intermediate guns should be bigger still, and so we built eight ships of the King Edward class which, in addition to the ordinary complement of four 1 2-inch and a number of 6-inch, carried four 9.2*3 in the upper deck. Japan just before the war with Russia ordered two somewhat similar ships with lo-inch guns as the intermediate armament.

Germany did not increase the size of her secondary guns, but she did design a ship (which was never built) known as the "S." The "S " was just like the ordinary battleship of the period except that instead of four 92's of the King Edward class, it was proposed to mount two extra big guns carried on either amidships, thus bringing the total of big guns up to six.

Meanwhile in the British Navy two ships of the Lord Nelson class were designed which, in addition to four 1 2-inch guns carried ten 9.2*8, and no 6-inch guns whatever.

At and about the same time a famous Italian naval architect suggested to his Government a ship which should be entirely armed with 1 2-inch guns, and nothing else, except some small guns for repelling torpedo attack. His argument was that such a ship instead of merely being able to disable an antagonist would blow her utterly to pieces.

The Italian Admiralty was pleased with the idea, but it did not adopt it on the grounds of expense. For that matter it suggested that England was the only country which could afford to build such a ship!

Cuniberti, like every inventor, was proud of his dream, and so it came about that in the 1903 "Fighting Ships" I published an article written by him describing this ideal battleship.

The immediate result was that I wished I had not, as everybody described the Cuniberti ideal ship as more suitable for the pages of H. G. Wells than for a serious publication dealing technically with matters naval. Now that Dreadnoughts exist in every navy of any importance whatever, I do not mind confessing that my own private opinion was about the same as that of the critics. And the real originator of Dreadnoughts is an English engineer, resident in Italy, called Charles de Grave Sells. He chanced to be a close personal friend of Cuniberti, and also a close personal friend of my own. To oblige Cuniberti he asked me to publish the idea, and to oblige him I did publish it! And there probably all three of us thought that the matter had ended.

Ended as like as not it would have, had it not been for the Russo-Japanese War. The Japanese being in for a life and death struggle for national existence, decided to lay down a couple of ships on Cuniberti lines. As a matter of fact these two ships, the Satsuma and Aki, never materialised as designed, as the Japanese could not depend on getting the necessary number of 12-inch guns. Consequently they only built a couple of ships which are always described as improved copies of the British Lord Nelson class, as they carry four 12-inch and twelve 10-inch in place of twelve 12-inch that they were originally designed for.

At and about the same time the United States contemplated the South Carolina, which carries eight 12-inch guns.

Then, and not till then, did the British Admiralty take action, although from that day to this it has been accused of having set the pace with a new kind of warship!

As a matter of fact the British Admiralty merely copied what other Powers were doing; but with a prescience which cannot be too highly commended, they saw to it that they came in first.

The first "all-big-gun-ship" to be launched and the first to be completed was the British Dreadnought, carrying ten 12-inch guns.

She was laid down in December, 1905. Three months later she was launched and eight months later she went to sea.

So far, so good. Unfortunately, just about that time the Navy was very much in the political arena; the Unionists accusing the Radicals of letting down the Navy, and the Radicals retaliated with the Dreadnought. The whole of it was human nature, and such blame as may be must be equally distributed. The net result, however, was a general concentration of world-attention on the Dreadnought and the building of Dreadnoughts by practically ever) Navy in existence. Today, something like a hundred have been completed, and very many more are completing.

Until the present general war broke out nobody knew for certain whether it was good policy to build ships costing two million pounds or more in place of ships which could be turned out for a million. Two million pounds is the minimum estimate of the cost of a Dreadnought. The Russian ones all come to nearly double that sum, and even some of ours though we build cheaper than anyone else, probably get well over the two million in actual fact. The real figures are generally so obscure that total cost and price per ton are merely forms of words to all save those technically interested. So far as the general public is concerned a pre-Dreadnought has usually cost about a million pounds, while a Dreadnought usually costs double.

The original Dreadnought carries ten 12-inch guns of which eight bear on one broadside.

Eight was selected for certain reasons along the theory that by a definite system of fire control if four guns miss, four others were absolutely certain to hit. Thereafter we subsequently rose to ten 12-inch guns, and then to ten 13.5-inch.

Theoretically, the 13.5-inch was selected on account of its greater penetrative power. As a matter of actual fact penetration of armour had nothing to do with the matter. The real story is that attempts to improve the 12-inch gun resulted in something so long that there was a waggle at the muzzle when it fired. This "waggle" was something so tiny that it could not be expressed in figures which any non-technical reader would understand. But at 10,000 yards (roughly, five miles) it meant that the long guns, though far more powerful, were less reliable than the older and shorter weapons in the matter of hitting.

Hence the 13.5, with which most of the ships knows as super-Dreadnoughts are armed. They are merely called super-Dreadnoughts on that account.

The 13-5's have proved themselves extremely reliable guns. They fire a heavier shot than the 12-inch, and being shorter are free from " waggle."

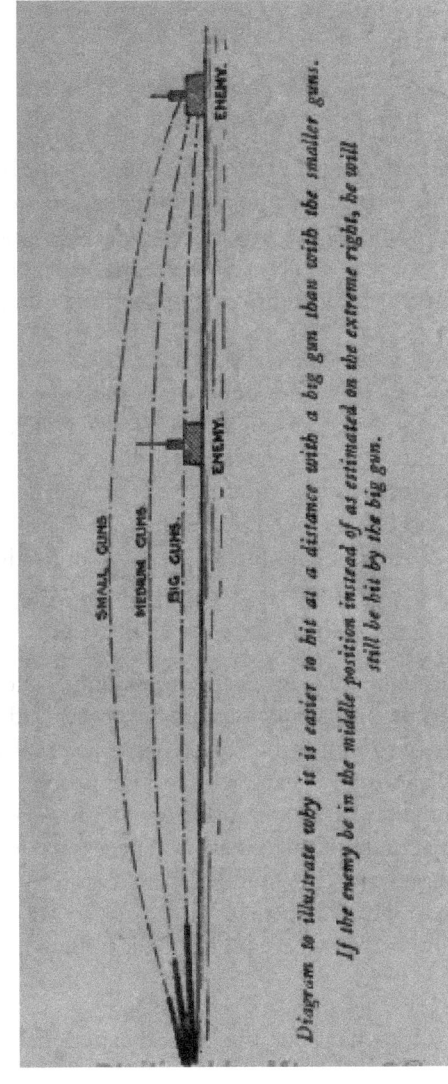

SMALL GUNS.

MEDIUM GUNS.

BIG GUNS.

ENEMY.

ENEMY.

Diagram to illustrate why it is easier to hit at a distance with a big gun than with the smaller guns.

If the enemy be in the middle position instead of as estimated on the extreme right, he will still be hit by the big gun.

The heavier the shot the easier the hitting. That is to say, an ordinary gunner could get nine shots out of ten on the target with a 13.5, where he would certainly not get more than eight, or perhaps even seven, out of ten with a long 12-inch; and though with the older 45 calibre 12-inch he might hit nearly as often, the damage from the lighter projectile would be infinitely less.

Hitting with light projectiles becomes problematical directly long ranges are required. At short range all firing is point blank; but the moment that the range runs into miles the delicate question of elevation comes into account, and the smaller the gun the sooner that question arises. Quite a trivial and microscopic error at the gun may mean several hundred yards error at the target. The bigger the gun the less the error.

Hence the general adoption of 15-inch guns for the newest big ships, although there is no armour in existence which the 12-inch gun is not able to penetrate at any rate theoretically. But the shooting with a 12-inch must be infinitely more careful than the shooting with a 15 -inch; hence the bigger gun. So much is this the case that it is probably easier to hit a ship with a 15 -inch gun five miles off than to hit the same ship at two miles with a 6-inch gun.

The diagram will explain the reason why. The lighter the projectile the more the calculation that has to be made about firing with it. With a really big gun like the 15-inch, gunnery gets reduced to something like shooting at haystacks and the problem of skill becomes mild. This is a matter of great importance, because, although in peace time it is comparatively easy to make the necessary adjustment of elevation for target practice, there is every reason to believe that in war the knowledge of what the penalty of error is so great that uncertain firing may result, except where point-blank firing comes into play.

Such at least has been the general experience in previous wars.

No 15 -inch guns will take part in the present war, unless it lasts considerably longer than anyone expects, as no ships mounting them can be completed for a good while to come. In the German Navy nothing bigger than the 1 2-inch exists in completed ships, but a great many British 13-5's will be in action, and we are likely enough to discover eventually the difference between the two guns which I have explained previously.

144

On their part the Germans have always believed that rapidity of fire is more important than accuracy, on the principle of the old proverb that if you throw enough mud some of it is certain to stick! In this there is a good deal of truth.

However, the British view is and always has been that the ship which gets in the first hit and can manage to keep on hitting with moderate regularity is bound to win and to receive no hits in return.

This brings us to the question of defence. As everyone knows, battleships, or "capital ships" as they are also called, are heavily protected with armour. The amount of armour protection varies from 25 to 33 per cent, or more of the total weight of the ship.

In the British Navy 25 per cent, in protection is rarely exceeded. In the German Navy the tendency is towards 30 or even 35 per cent, expended in protection.

The particular problem involved is at least fifty years old. On the one side distinguished people have argued that everything should be sacrificed to killing the enemy and hitting him so hard that he cannot hit back.

On the other hand equally distinguished people have argued in the past that to hit the enemy fairly and continually it is necessary so far as possible to be so armoured that you yourself stand little risk of being badly hurt and so prevented from hitting.

From that day to this no one has ever ascertained which view is the right one. By a curious freak of chance in all fighting between battleships which has taken place, the lightly armoured ships have been hit badly, while the more heavily armoured ones have never been hit at all.

The present war may solve these problems for once and all, but again chance may intervene and no reliable data be secured.

Be that as it may, however, the British Navy as a general rule is committed to the general principle of "hit first and take your chance about defence." The German, on the other hand, is as heavily committed to an opposite principle, the ability to stand hard hitting at the sacrifice of ability to hit hard.

The British system seems the better one but, as remarked previously, both ideas are a matter of theory which a few months ago only a very few people expected would ever be put to the test.

II. THE BATTLE CRUISER.

The battle cruiser is essentially a ship of the Dreadnought era. The title "cruiser" is altogether a misnomer as they are not intended for cruiser work.

Battle cruisers vary in nature and in every degree, but the general principles of all of them are the same, that is to say they carry fewer guns than a Dreadnought battleship, have thinner armour, are larger and heavier, and a great deal faster.

For example, the battle cruiser Lion is 26,350 tons, compared to the 23,000 of the King George V. She is 100 feet longer and of 75,000 h.p., against 28,000. Her speed is in the region of 28 knots as compared to the 21 knots or so of the King George V. The speeds of battle cruisers are, however, uncertain quantities; that is to say there is always a great deal of difference between what they are supposed to do and what they actually can do. All of them are capable of short spurts considerably in excess of what they develop in the normal way, or for long periods.

The cost of building a battle cruiser is about the same as that of a battleship that is to say, about two million pounds. The cost of upkeep is, however, a great deal more, as with its immense engines the coal consumption is enormous. Burning coal only, the full consumption of coal of a big battle cruiser like the Lion is very little short of 1,000 tons a day. As a rule oil is burned in addition to coal. This reduces the coal consumption, but as oil costs about £4 a ton, no economy is effected thereby. The reason oil is used in conjunction with coal is because of the greater efficiency thereby obtained.

The total amount of coal carried by a ship like the Lion is 3,500 tons, in addition to which there are 1,000 tons of oil. From this it will be seen that were a battle cruiser always steaming at full speed, the enormous quantity of fuel which she carries would last her only a few days. As a matter of fact, however, no warships are ever steaming at full speed like the Atlantic liners do. There is a great difference between warships and merchant vessels, and it is this which makes the work of the engineer in the Navy harder than in the mercantile marine. It is much easier to maintain a constant speed even though it may be a high one, than to be continually chopping and changing, as has to be done with a warship.

The war duties of a battle cruiser are essentially of a destructive nature. Should the enemy's fleet be sighted, the battle cruisers, as a fast wing, have to go in pursuit and bring it into action. How this is brought about is best explained by the diagram on the previous page. From this it will be seen that a couple of battle cruisers can easily be a tremendous menace to an extremely powerful fleet, as coming up from astern they can concentrate on the last ship. Should the enemy go on, its last ships will be disabled in detail. If, on the other hand the enemy stops to fight the battle cruisers, the object of the battle cruisers is achieved, for the enemy will then be caught up with and compelled to fight with the pursuing main fleet.

The duties of battle cruisers being as they are, great care is exercised in selecting the most dashing and daring officers to command them.

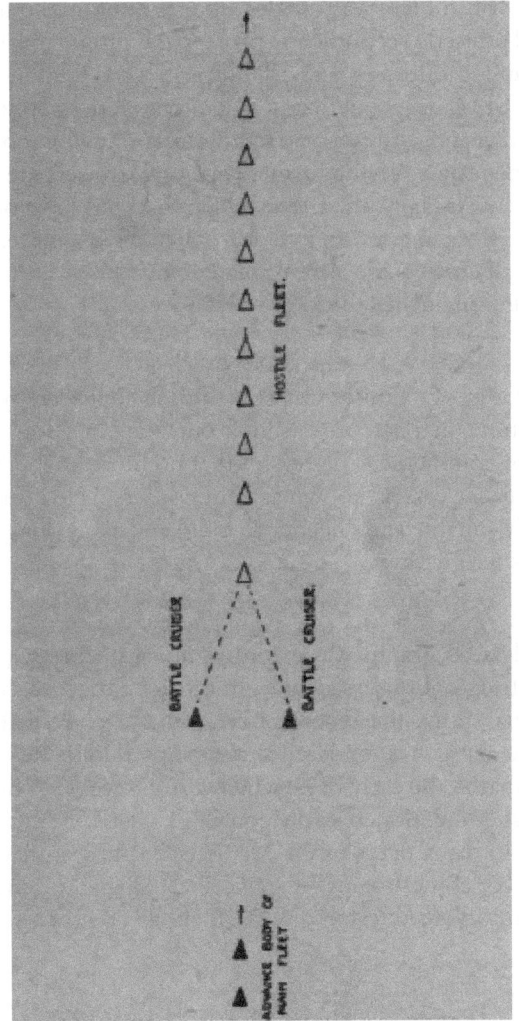

Illustrating how Battle Cruisers act against a battle fleet

III. CRUISERS.

Time was when the British Fleet was divided up into armoured cruisers, first- class protected cruisers, second-class protected cruisers, third-class protected cruisers, un-armoured cruisers, lightly armoured cruisers and scouts. But a couple of years ago all these distinctions were swept away in favour of two main divisions cruisers and light cruisers; anything over 6,000 tons displacement being in the former category. Cruisers are no longer built for the British Navy, the theory being that they have outlived their period of utility. Save in a few old vessels which have long ceased to have any war utility, all cruisers are armour plated, their belts in the majority of cases being six inches thick.

The original idea of a cruiser was some- thing equivalent to the frigates of Nelson's time, that is to say, as look-out ships, and for the attack and defence of commerce. The reason that no more are built is partly because the smaller vessels known as light cruisers have been found more generally suitable and cheaper to build and also owing to the invention of wireless telegraphy.

Before wireless telegraphy was invented, it was necessary (see diagram) to have a string of cruisers all in sight of each other, so that messages by flag signal or masthead semaphore could be conveyed along the line. Wireless telegraphy has rendered all that superfluous. It is interesting to recall that in the early days, when wireless was first heard of, Admiral Lord Fisher mentioned in the course of a series of lectures which he was giving to the officers of the Mediterranean Fleet that " if wireless telegraphy ever comes to anything, we shall be able to scrap half our cruisers, and in future need to build fewer than we do now."

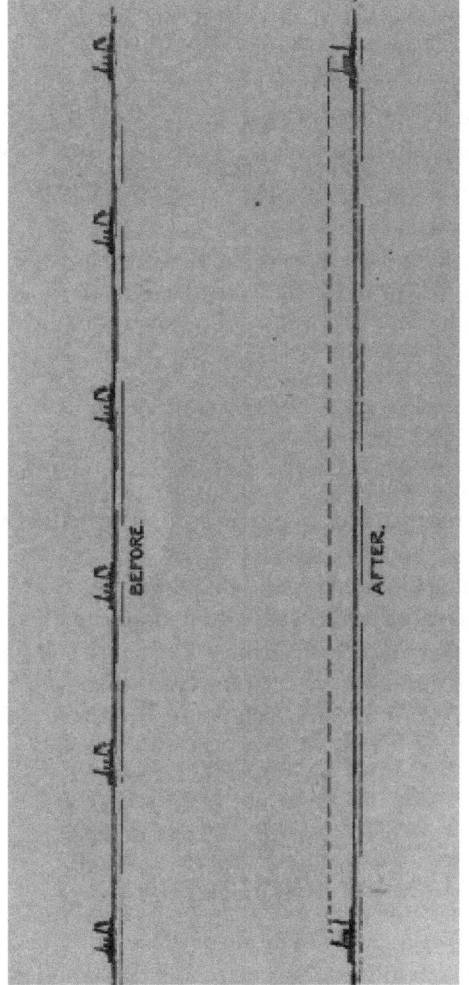

Diagram to Illustrate the savings in Cruisers Brought about by Wireless Telegraphy

The words were prophetic. It is not to be imagined that our armoured cruisers are useless. Far from it. They still remain the cavalry of the seas, and act as screens between hostile fleets just as mounted soldiers do between hostile armies.

Here again matters are best made clear to the uninitiated by means of a diagram. Supposing that Fleet A is looking for Fleet B and Fleet B does not wish to be found. B protects himself by putting out his cruisers. Consequently to locate B, A's cruisers must break through B's cruisers, it being obviously impossible for the battle fleet itself to split up searching for the enemy, who would then get it at a disadvantage. It was on account

of these screens that the cruisers came to be armoured, for before wireless came into use not only did they have to fight their way through, but what was still more important, fight their way back again. It was also necessary to have them very numerous, so that they could see and support each other.

Another most important duty of cruisers also a duty which has of late years become much more urgent, is the protection of battle fleets from torpedo craft. This is done by sweeping in the daytime up to a hundred miles or more ahead of and around the Fleet. This is a task which destroyers were invented to do against torpedo boats. It was soon found however, that the destroyer was capable of being used as a much more efficient and deadly torpedo boat than the craft which she was designed to destroy. So it is destroyers which the cruisers have to hunt for in the daytime.

At first sight it would look as though cruisers capable of only 22 or 23 knots could not do much in the way of catching destroyers of 30 knots or more. As a matter of fact, however, destroyers are very easily caught, as the sea is rarely calm, and the moment the weather becomes at all rough the destroyer which in smooth water could do 30 knots will find her speed reduced to 20 knots or less.

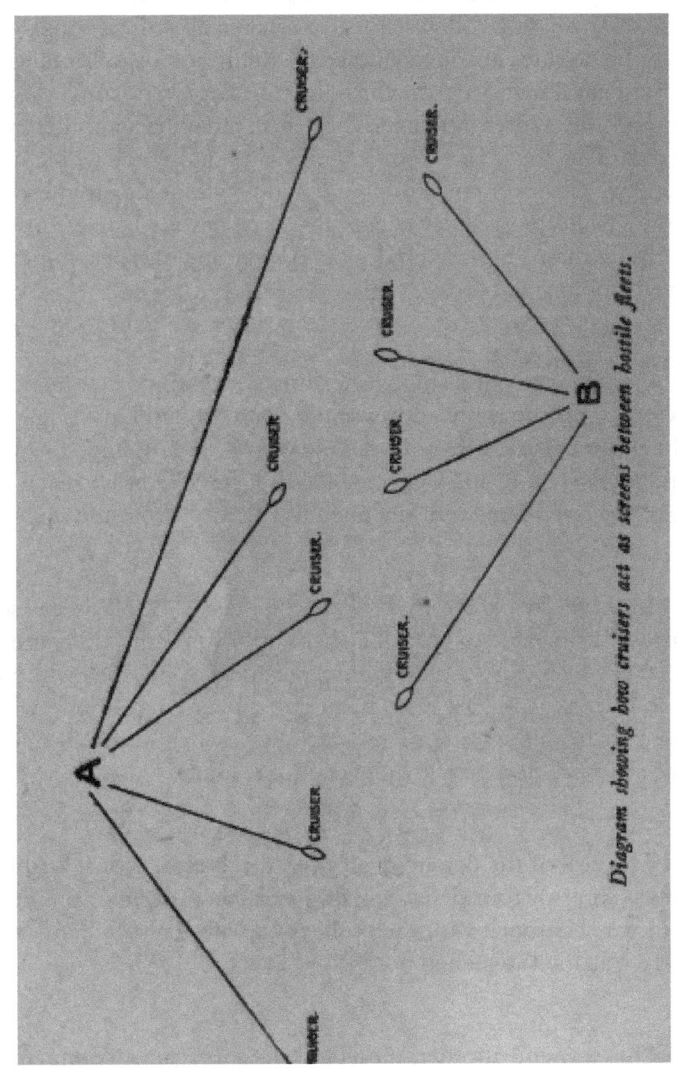

Diagram showing how cruisers act as screens between hostile fleets.

In addition no destroyer can carry enough fuel to stand being chased all day and then to return again and attack the battle fleet in the night.

The duties of a cruiser engaged in commerce warfare vary a great deal according to whether she is attacking or defending it. The whole of the trade of any nation follows along a certain well-known route, outside of which the ocean is absolutely deserted. Many years ago I was with a fleet during naval manoeuvres which, leaving Bantry, steamed slowly to the south of the Azores, remained there for some time, and then returned to England. The cruise lasted three weeks, and during the whole of the time no news whatever concerning it could be gleaned from merchant ships. We saw nothing all day except the same deserted wastes of water. Yet not more than a hundred miles or so away any number of merchant ships were coming and going.

Consequently, if a ship wishes to attack commerce, she must do so by keeping off the route and swooping on to it and capturing something as occasion offers. In the old days of sailing ships this was a busy and lucrative job, and our trade losses were very heavy, even after Trafalgar when we had swept the main hostile fleet from the seas.

In these days of steam, however, the task is by no means so easy, as instead of being able to keep the chase up for months, there are very few cruisers which can keep up commerce attack for a week without having to re-coal. And so it is no longer possible to make those wide swoops away from the trade route which now, when nearly all ships carry wireless, are more necessary than ever for the attack.

The task of the defending cruiser is, therefore, considerably simplified. Three or four defending cruisers scattered about on the route, and all in wireless touch with each other and with the merchant ships, can easily arrive at any threatened port.

One way and another, therefore, the war against commerce has lost most of its glamour, and it is likely to cost the attackers a good deal more than they get out of it, especially as even if they do capture vessels, there remains the problem of getting them into port.

IV. TORPEDO CRAFT.

As mentioned in an earlier chapter, destroyers were originally designed to destroy torpedo boats ; but it was soon found out that they were so effective as torpedo boats themselves, that now they exist for that special purpose, and torpedo boats proper are rarely built.

Torpedo boats first appeared in the war between Russia and Turkey in 1877. They were nothing more than steam launches of the old type small craft with a long spar fitted in the bow. At the end of the pole was some gun-cotton or other explosive, and the idea was to charge a ship and then fire the explosive. In practice this occasionally did some slight damage to the ship, but usually destroyed the boat entirely.

The Russians utilised such boats for carrying torpedoes, which were generally slung underneath the keel, and let go as required. Later on a special gear was fitted for dropping the torpedoes.

Torpedo boats were, however, of very small use until torpedo tubes were invented. The torpedo tube is a species of gun which fires the torpedo into the water. Once in the water the torpedo sets itself going by its own mechanism (see a later chapter on Torpedoes).

About twenty years ago torpedo boats had increased in size up to seventy or eighty tons. They had long been considered a danger, and some gunboats officially known as torpedo-boat catchers, but always spoken of as "catchers " simply, were in existence. These gunboats varied from 400 up to 1,000 tons, but they were never large enough to be able to steam properly in a seaway. Experience in manoeuvres showed that the torpedo boats generally escaped from them quite easily.

For some time the Germans in addition to building torpedo boats, had been in the habit of building some larger vessels which they called " division boats." These were a good deal faster and larger than the ordinary torpedo boats which they were intended to lead into action.

It was probably these German boats which first gave birth to the idea of destroyers. The first two ever built were the Havock and Hornet. They were of some where about 270 tons displacement, and capable of a speed of 27 knots.

When first designed they were alluded to as "catchers," but that being a name of unfortunate memory, some genius hit on the idea of calling them "torpedo boat destroyers," and from that day to this that has been their official name, abbreviated into " t.b.d."

The first two. destroyers took the sea in 1894, and were sent into the Bay of Biscay to note how they fared. There was considerable divergence of opinion as to what would happen to them ; they were not considered safe to go out by themselves, and the " catcher " Seagull was sent to look after them.

Beyond the fact that all on board got very seasick, the destroyers came through the test with complete success, and the Admiralty immediately ordered a very large number of them, and has gone on building them ever since.

The functions of destroyers are obvious. Their attacks are, of course, made by night, and the danger of them is increased by the fact that they act in divisions.

The present war will prove whether or no destroyers are as dangerous as they have been reckoned to be against big ships.

The past history of destroyers has been rather against the torpedo that is to say, in the past it has never done what was expected of it. In the war between China and Japan the torpedo boats did nothing in the open sea. On the other hand, they delivered an attack on the Chinese Fleet anchored in Wei-hei-Wei with very considerable success, in conditions which were all against them.

In the war between Spain and America, nothing was done by torpedo boats.

Diagrams showing proportionate sizes of warships.

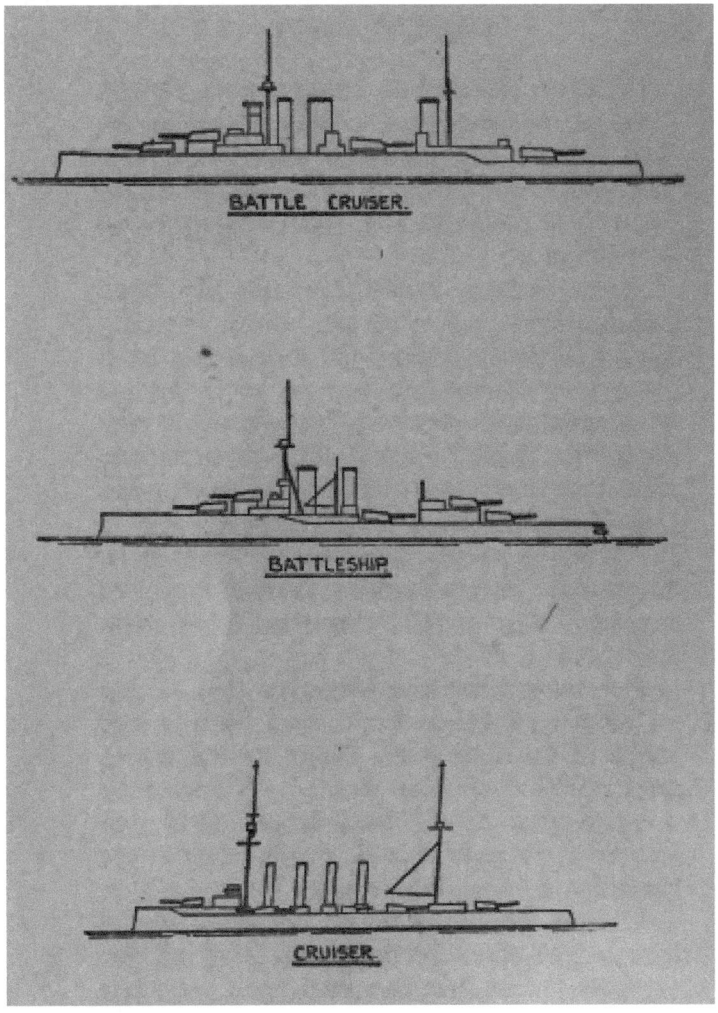

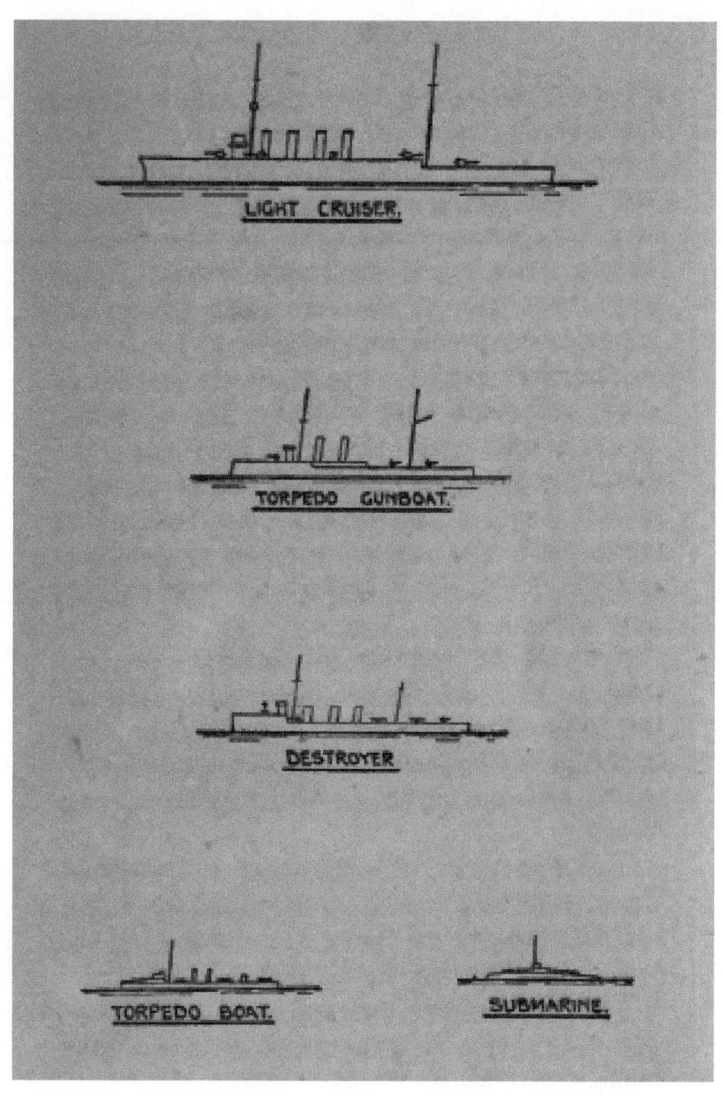

In the more recent war between Russia and Japan a surprise destroyer attack was delivered by the Japanese on the Russian Fleet at Port Arthur. Although a great many boats attacked, and although the surprise was complete and the Russian response amounted to nothing, only three ships were hit, and the damage done to each of these was in every case very slight.

Thereafter, destroyers did nothing on either side beyond occasional fights with each other until the great Naval Battle of Tsushima. The night following this battle several of the disabled Russian ships were sunk by torpedo attack, but as most of them must have been in a sinking condition already, the loss was held to prove very little one way or the other.

In the war between Turkey and Greece the Turkish cruiser Hamidieh was torpedoed, but a month later she was repaired, and making her famous cruises.

One or two old Turkish ironclads of prehistoric date were torpedoed by the Greeks, but it cannot truthfully be said that the torpedo influenced the war in any way whatever.

One reason for this is that each side has always had so many torpedo craft that they have neutralised each other a condition of affairs which is likely enough to obtain in the present war also. Another reason for the ill-success of torpedo craft, except in attacks on ships in harbour, is that the sea is a vast place. When destroyers look for battleships they are somewhat in the position of the man looking for the needle in the proverbial bundle of hay. If he finds it, he will secure it ; but all the odds are against the necessary finding.

This is the defence upon which battle fleets mostly rely. Their other defence consists, on paper, in the special guns which all ships carry to destroy attacking torpedo craft. Until the present war nobody had any faith in these little guns, on the grounds that although a destroyer might be badly hit, she would probably ease off her torpedoes before she sank.

Since the sinking of the German Kzenigen Luise by a British destroyer with four shots, the small gun is more esteemed than it used to

be, but the only real direct defence of big ships against torpedo attack is
still held to be the big gun.

V. SUBMARINES.

We are accustomed to think of the submarine as an entirely modern invention. In one way it is, but 300 odd years before Christ, Alexander the Great had his own ideas on the subject. Coming to more recent but still remote times, our own King James the First took a trip in a submarine invented by his Court Physician.

In the American Civil War of fifty years ago, things were done by submarines. Usually, or almost invariably it was at the expense of the submarines. Still the attempts were made.

The trouble with submarines was the provision of adequate motive power while they were under water. It is only quite in our own times that considerable power has been developed for under-water propulsion. This is obtained electrically by accumulators.

Here it should be noted that the popular idea that a submarine is a boat which works under water most of its time is entirely incorrect. What a submarine actually does is to work on the surface. She only goes under water for the actual purpose of attacking the enemy, or to avoid being attacked by some above-water craft, which is looking for her.

For this surface propulsion submarines use internal combustion engines, or an adaptation of ordinary steam engines steam at the present moment being rather more in favour than the modified motor-car engine. The Diesel engine although all right for small powers, has so far not proved very satisfactory directly anything over 1,000 h.p. is required, and the modern submarine requires a great deal more power than that.

To an extent which few people realise, the submarine has grown tremendously during the last few years.

The original idea of it was a little boat in which a couple of men risked their lives in an endeavour to destroy the enemy.

To-day the modern submarine carries quite a considerable crew, and is as big or bigger than a destroyer. It has sleeping accommodation, and is generally speaking a self-contained ship something altogether different to the elementary boats in which the crew had to sit down and dare not move for fear of upsetting the whole thing.

The first time that real submarines were actually used in warfare was in the Russo-Japanese War of about ten years ago. The Russians had only one submarine engaged. This was a little one-man affair which started out from Port Arthur with a view to sinking Admiral Togo's fleet. The next heard of it was after the war, when it was dredged up by the Japanese salving sunken Russian ships.

In the same year the Japanese employed five submarines. They were worked by men who necessarily knew nothing about such novel weapons, and all five of them sank and were not recovered until the war was over.

In the more recent Turko-Greek War the Turks had no submarines, and the Greeks had only one available and made no particular use of it.

As a result the present is the first war in which submarines have taken part as a regular arm. When the war started every-body had theories of what they might accomplish, and these theories ran the whole gamut of from nothing to everything. But nobody actually knew anything whatever.

The satisfactory side of the business is that whereas the British Navy started the war with something like seventy effective boats, the Germans had only half that number.

Chance, however, so willed it that the first demonstration was given by Germany. The stories written about H.M.S. Birmingham and the German submarine U 15, which was sunk, are, of course, entirely fictitious, except in so far that the Birmingham did sink U 15, but the real truth of the matter is that the U15 fired at a certain British ship and missed her. Thereafter the U15 might have got home in safety, had not her captain imagined that he had succeeded, and come to the surface to shout "Deutschland uber alles." That little incident settled the fate of U 15, as she came up alongside the Birmingham, and was sunk at once.

In the present stage of development, supposing a submarine to find a ship, the odds are very heavy that she hits her. It is practically impossible for any ship to detect an attacking submarine, or to know that she is attacked until the torpedo is fired. Once the torpedo is fired the game is given away, because submarine attacks can only be delivered in daylight, and torpedoes cannot be fired without betraying the circumstances by the bubbles that they make on the water.

To the ordinary public bubbles on the water might not convey any particular meaning; but to the Naval man they convey everything that matters. When ships are in company, as they usually are, the submarine that fires her torpedo and misses, has probably fired her last shot.

The penalty of failure is death for all concerned. In this, submarine warfare differs from other warfare, in which, however many may be killed, a certain number of the crew are more or less sure to survive. In submarines it is "all or none."

It is very difficult to give a non-technical definition of a submarine. In the ordinary way it floats in the water just like any other ship. By introducing water into certain tanks it reduces itself to the same specific gravity as the water, in which consequently it rises or sinks according to the action imparted to certain little paddles on its sides which are known as hydroplanes.

When submerged its course is steered by means of an instrument known as a periscope.

This is a kind of tube with mirrors in it whereby what is happening on the surface is reflected below.

A submarine could not stay under the water for any great length of time, because the air would become vitiated. This condition of affairs is met by various appliances for purifying the atmosphere and adding oxygen to it.

The most modern submarines are capable of remaining under water for twenty-four hours if necessary, but the ordeal is naturally a trying one.

VI. (a) AIRSHIPS.

The Naval Air Service is responsible for all the dirigibles and has in addition its own force of aeroplanes.

Dirigibles are of various sizes, but there is nothing in the British Navy to correspond with the enormous German Zeppelins construction in England being in its infancy. Indeed, the only two effective dirigibles in service at the outbreak of the war were foreign, one being made in Germany, and the other of French construction. Reference to these will be made later.

It is necessary first to explain what a dirigible airship is. There are two classes, which differ a great deal from each other except that of course both are filled with hydrogen gas and depend for flotation on the fact that they are lighter than air.

Battle airships like the German Zeppelins are characterised by a rigid framework, usually of metal, but sometimes of wood. They are divided into a number of compartments, usually from fifteen to seventeen.

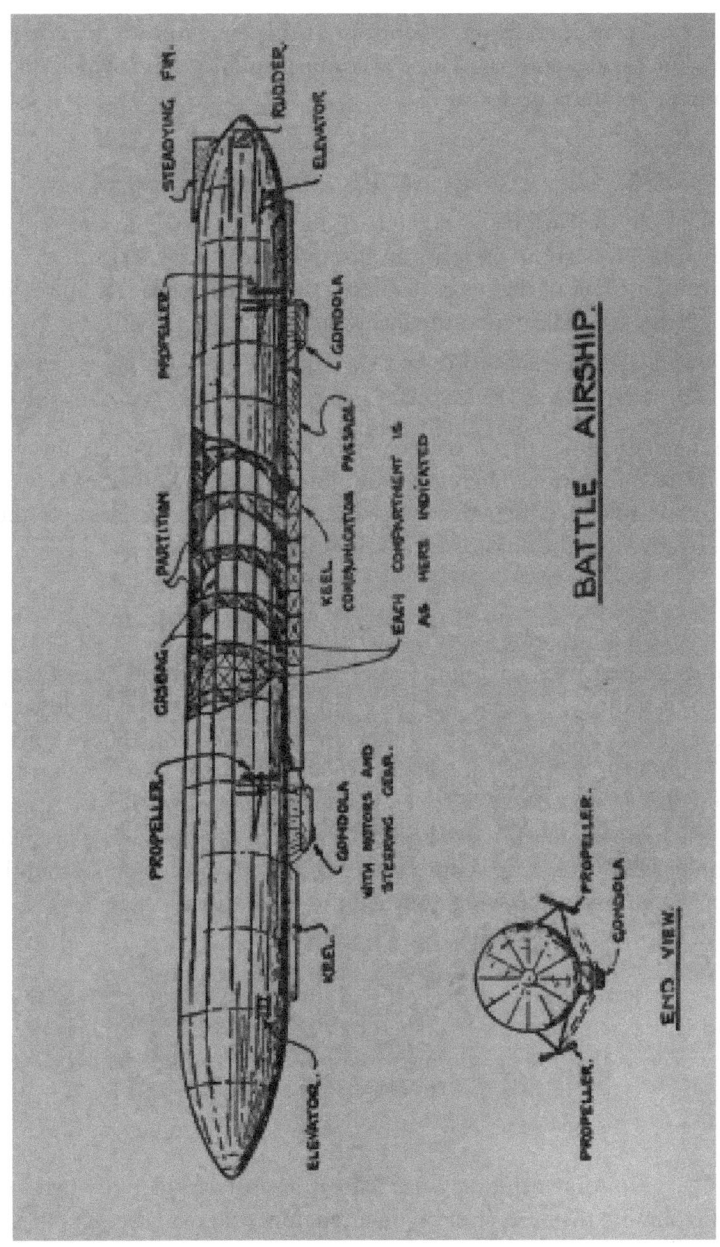

STEADYING FIN.

RUDDER.

ELEVATOR.

PROPELLER.

GONDOLA.

PARTITION

KEEL COMMUNICATION PASSAGE

EACH COMPARTMENT IS AS HERE INDICATED

GASBAG.

PROPELLER.

GONDOLA WITH MOTORS AND STEERING GEAR.

KEEL.

ELEVATOR.

BATTLE AIRSHIP.

END VIEW.

PROPELLER.

GONDOLA

PROPELLER.

Each compartment contains a gas-bag which is in substance a separate drum-shaped balloon. This group of balloons is surrounded by a large outer covering known as the envelope.

Suspended from the bottom of this envelope are two or three cars known as gondolas. These accommodate the crew, which may number anything up to thirty.

The engines are usually three in number, developing a total of 450 h.p. in the earlier and 800 h.p. in the very latest. The speed of the early ones was about 35 miles an hour, whereas the latest run to over 50. They do not work the propellers directly, as in aeroplanes, but by means of gearing, as indicated in the diagram.

No two large dirigibles are exactly alike in their steering arrangements, but the principle whereby they rise and fall is always like that of a submarine. There are a number of planes known as elevators, corresponding to the hydroplanes of a submarine. According to the way these are tilted the ship rises or descends.

In addition to bombs the battle airship is capable of carrying guns, the Germans having mounted as many as four on one of theirs. Two of these were carried in the gondolas, one could be let down well clear of everything, while the other was placed on top. These guns are for defence against aeroplanes. The particular German ship which was thus armed is supposed to have been lost early in the war owing to attempting to use her top gun. It is assumed that there was a leakage of gas which would, of course, escape upwards, and that firing the gun ignited this gas, causing an explosion which burned the whole airship.

Battle airships at the present time are extremely fragile things, and very open to attack from aeroplanes. The reason of their use, however, is that they can cruise about at night and keep up for a long time.

Not only can they carry a great deal of petrol, but when the wind is favourable they can shut off their engines and drift, becoming a species of sailing ship.

The war duties of a battle airship are to scout ahead of the fleet and drop bombs on hostile vessels. The damage to be done to big ships in this way is not very great, but the menace to small craft and submarines is quite another matter. However, hitting small, moving objects is by no means easy.

For defence against airships warships carry special guns capable of firing fire-shell high into the air. A battle airship is somewhere about the size of a Dreadnought, i.e. 500 feet, and this adds to the inconvenience of them, for immense sheds have to be constructed to contain them. For this reason it is only with the Germans that the battle airship has been popular. The French have only one. Four or five years ago, one was specially built for the British Navy at Barrow. She took a very long time to build and came to grief shortly after emerging from her shed.

Small airships are known as non-rigids because they have no frame work, but consist of a single envelope inflated with gas. The advantage of this type is that they can be packed up and sent about on board ship or transported overland in wagons.

The usual method of rising and falling is to make use of two bags known as ballonets, into which air is pumped, thus altering the

trim. The larger and more modern of these air cruisers are generally divided into two or three compartments. The Astra-Torres (French make) is peculiarly divided so that she has the form of one sausage placed on top of two others. The endurance of these small cruisers is, of course, much less than that of a battle airship, and so their war work consists in inshore scouting.

The average crew is about five.

There are also some very small airships used for instructional purposes, of which the principal and interesting feature is that the propellers are mounted on a swivel, and so in addition to being used for going ahead or astern are valuable for rising and falling.

(b) AEROPLANES.

Aeroplanes differ from airships in that they are heavier than air and keep up owing to the speed at which they travel. There is no essential difference between an aeroplane, a seaplane or a flying boat. The general principle of all is the same. The land aeroplane is, as everyone now knows, mounted on wheels, on which it runs along until it rises. With a seaplane, floats are substituted and the machine runs along the water until it rises.

In flying boats the principle is the same, but instead of floats there is a species of canoe.

Aeroplanes required for naval purposes have more powerful engines than land machines, because the construction of them has to be a great deal stronger and heavier.

The usual crew of a seaplane is two men, of whom one does the driving, while the other acts as observer, or bomb dropper, as the case may be. The probable war uses of a seaplane are scouting around the coast, or accompanying the fleet in a special mother-ship vessel, from which they can be despatched as required.

War aeroplanes can be divided into three classes: first, very fast one-man machines for special work. A machine of this sort would carry no bombs and generally as little superfluous weight as possible, everything being sacrificed to speed. The nearest land analogy is the motor-bicycle messenger.

Next come the intermediate machines, intended for general observation purposes. Such machines might carry a bomb or two in addition to its two men, but would not be otherwise armed, and it would require sufficient petrol so as to give the maximum endurance. Such machines are generally designed to be fitted with wireless telegraphy for the transmission of information.

Thirdly, there is the war aeroplane a much bigger affair than the other two. Its vital parts are generally protected with bullet-proof armour.

It carries a special gun, usually mounted forward, and its war duty is to attack hostile airships or aeroplanes which may be defending them.

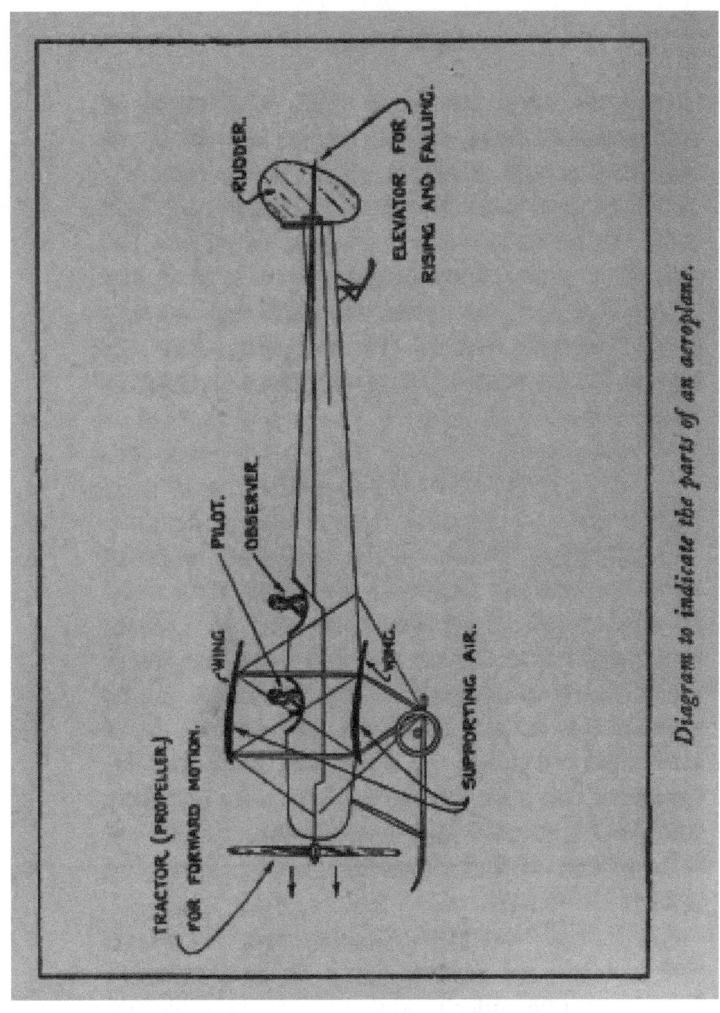

Diagram to indicate the parts of an aeroplane.

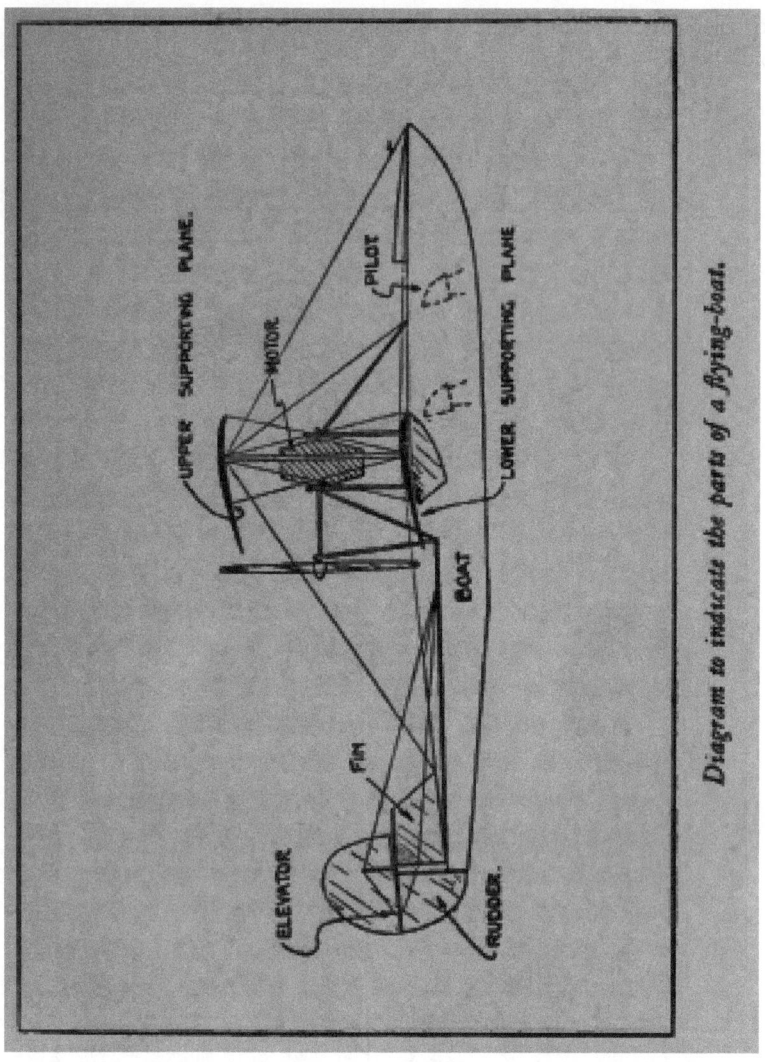

Diagram to indicate the parts of a flying-boat.

It must be understood that the diagrams illustrating aircraft are of an entirely general nature. Some aeroplanes have the screw in front, when it is known as a "tractor," because it drags the aeroplane after it through the air. Others have it placed behind, where it is known as a "pusher," because it pushes the machine through the air. The position of the screw is entirely a matter of general convenience, and the difference is merely that of pushing or pulling a wheel-barrow.

VII. MINES.

The mine is a very old invention, being nothing but a development of that "petard " with which the engineer was hoisted as mentioned by Shakespeare. It was first developed along modern lines by the Russians at the time of the Crimean War in the Baltic, in 1854. They dropped about in the Baltic a good many of what were then known as "infernal machines." These were filled with seventy pounds of gunpowder, and exploded on contact. They did not do much damage to ships which struck them, but several of our people were injured while examining unexploded mines which they had picked up.

Further developments took place in the American Civil War, the Confederates being considerably active in strewing their rivers with mines which were then called "torpedoes," a name they retained until the Whitehead automobile torpedo came to be known by the title of "torpedo" the old original torpedo then being named "submarine mine."

Till within a few years ago the favourite form of mine was a systematic harbour defence laid by the Royal Engineers. It consisted of a large number of mines connected to the shore with a cable and capable of being fired electrically at any moment that an enemy's ship or fleet was over it.

Since the coming of the submarine the old style of mine field has more or less fallen into disuse, but there are still some of them about, and they are liable to exist in very unexpected places.

The mines of which we have heard so much ever since the Russo-Japanese War are a direct adaptation of the old Russian "infernal machine." That is to say, they are intended to float and explode on contact with a ship.

They are generally known as "blockade mines," because their main use is to lay them outside a harbour in which a fleet is blockaded or shut in. Mines of this sort remain in the position where they are submerged, floating just below the surface of the water, invisible to the enemy, but of course, well known to the layers. Occasionally such mines have broken loose and damage has resulted there from. But that has been due to faulty construction. A properly-built mine will, if it breaks adrift, automatically sink, or otherwise become innocuous.

It is against the laws and customs of nations to lay mines out in the open sea, as the Germans have done, as such mines are dangerous to all neutral shipping, and unlikely to inflict damage on the enemy, except by chance.

There are one or two ships specially built for mine-laying the French, German and Russian Navies having built a couple each; but the general rule, and that followed in the British Navy, is to adapt for the purpose old cruisers who have outlived their sphere of war utility.

With a view to saving weight and space, most or all of the guns are removed, and a large hole or holes cut in the stern. The mines are stowed in the middle of the ship, and run along a small railway to the stern, whence they are dropped into the sea at definite intervals, these intervals being such that it is not possible for a ship to pass between the gaps. Usually, they are fastened together in pairs, this being the general practice in the Russo-Japanese War.

When the enemy is suspected of having laid mines in any particular place, the first course is to remove them. For this purpose small vessels of little value are employed, steam trawlers in particular being favoured for the purpose. The work, of course, is not unattended with danger, everything depending on the organisation and methods adopted.

The less said about this matter, however, the better, beyond remarking that the side which lays mines is very likely to find its own mines taken up and used against it.

Indeed it is said that the mines which sank two Japanese battleships off Port Arthur were ones which had been originally laid by Admiral Togo in some other spot. The Russians having kept observation, shifted the mines into the route which the Japanese ships were in the habit of following.

A similar story has also been current to the effect that it was a Russian mine shifted by the Japanese which sank the Russian flagship Petropavlosk.

There is nothing improbable in either story. It shows the care which has to be exercised by those who employ this form of warfare.

VIII. PERSONNEL: OFFICERS.

The present system of officering the British Navy is as follows: Boys who wish to become Naval officers are examined at about the age of twelve by a special Board. There is no rule-of- thumb examination which can be crammed for; the whole idea is to select from the candidates such boys as promise to be resourceful, or who exhibit ability to think.

Many of us know the famous picture in "Punch" of an admiral examining candidates for the Navy. The question asked by the examiner was " Name the two greatest British Admirals." The boy is represented to have replied "Nelson, and please, sir, I have forgotten what your name is!"

There is some reason to believe that this incident actually occurred; but it is more than doubtful whether that boy is a naval officer at the present moment!

Another question which has been asked of would-be embryo officers is : " If one train leaves York at 50 miles an hour, and another leaves London at 30 miles an hour, which will be nearest to London when the two meet ? " This old " catch " is a very favourite question or rather used to be a favourite question, as crammers have now pointed out that when two trains are going in the opposite directions, when both meet they must naturally be exactly the same distance from London!

However, for a long time it was the chief question, and the devising of a better one to discover intelligence takes a fairly long time.

This system of selecting our future officers has been strongly criticised in certain directions ; but when all is said and done it is difficult to think of anything really better. A system whereby boys are crammed to answer certain questions on subjects which are known beforehand, cannot possibly be a test of adaptability, and adaptability is the first essential for a naval officer.

It may further be added that the Examination Board is continually changing, so that anything in the nature of favouritism is impossible; while as the fees charged are much less than those of any big public school, and very little if anything in advance of the charges of an ordinary private school, it will be seen that "class interest "plays little part in the selection of our future officers.

There is, it is true, a preference for the sons of existing officers, but that is how things should be. "Comes of fighting stock" still means as much for men as it does for dogs. For the rest, it may be taken for granted that at the present day the Navy gets its pick of the best available stuff.

When a boy is selected he is sent to Osborne Naval College, from which he eventually emerges as a naval cadet, and then as midshipman, with the pay of £31 18s. 9d. a year, from which £5 is deducted for his instruction.

From this stage he proceeds to acting sub-lieutenant at twice the amount, and eventually becomes a sub-lieutenant on 5s (JC s= shillings) a day.

Thereafter his career depends entirely on himself. According to how he works when he is an acting sub-lieutenant his ultimate future absolutely depends.

When finally promoted to lieutenant he receives 5s. a day, plus various allowances, which can bring his pay up to 14s. a day or more.

After four years' service he automatically rises an extra 1s. a day, after six years 2s a day, after eight years 3s. (when he automatically becomes a lieutenant-commander), after ten years 43., after twelve years 55., and after 14 years 6s. If, however, he has remained a lieutenant, or lieutenant-commander for so long as this the end of his career is reached.

As a lieutenant his duties are as follows: should he have "specialised," i.e., done well enough in examinations as a junior to be entered for a special course in gunnery or torpedo engineering or navigation, his principal duties on shipboard as a lieutenant will lie in one or other of those directions.

His main job will be looking after the guns and the fire control as "Gunnery Jack" ; the torpedoes and all the electrical gear of the ship, from searchlights to electric bells (as " Torps," or " Torpedo Jack") ; or if a navigator (" the Pilot ") on him will rest the responsibility that the ship's course is correctly piloted. Should the ship ever get aground, he will have finished his service career the Navy knows no mercy in matters of this sort.

A non-specialist officer (known colloquially as "salt horse ") serves as a watch- keeper. His duties are to be on the bridge during the "watch."

A "watch" consists of four hours' duty: midnight 4 a.m. (middle watch); 4 a.m. 8 a.m. (morning watch); 8 a.m. noon (forenoon watch); noon 4 p.m. (afternoon watch) ; 4 p.m. 6 p.m. (first dog watch);

6 p.m. 8 p.m. (second dog watch).

A "dog watch," it should be explained, was invented so that no man should consecutively keep the trying middle watch two days running.

The next stage of promotion in the career of a naval officer is "commander."

The pay of a commander is 22s. a day, which is just a little over £400 a year. In addition there are various extras which he can expect to draw.

But on the other hand, if he aspires to further promotion he will certainly have to put his hand deep into his pocket. The smartness and efficiency of a ship depends upon the commander, and whatever may obtain in theory, the creation of a "smart ship " means spending money in various ways. In addition to this, should he be selected for promotion to captain, he is likely to remain some considerable time on the half-pay list, and the pay of an unemployed captain is not high.

When first employed as a captain his pay may be as much as £700 a year, and eventually one way and another he will, if employed arrive in the neighbourhood of a thousand a year, out of which he will have to meet various heavy expenses.

If promoted to commodore which is a sort of cross between a captain and an admiral he will receive £1,000 a year, plus extras. As a rear-admiral, he will receive the same, his extras will be the same, but his expenses considerably heavier, as the entertaining of all and sundry will fall upon his pocket. In addition thereunto, he will have a considerable period of non-employment on half-pay.

The career of an officer, once he has reached the rank of captain, is automatic thereafter. His promotion depends entirely on seniority. His employment that is to say, being on full pay instead of half-pay, depends entirely on his abilities.

On promotion to vice-admiral, our officer, if employed, will receive £4 a day, which amounts to something like £1,500 a year. In addition to this there are various allowances ; but if he actually makes 500 a year clear out of his employment he will deem himself lucky.

An admiral receives £5 a day, and in addition various allowances. By the time all his expenses are settled up, his actual receipts amount to something rather less than those of a London motor-bus driver, although in the eyes of everyone he is earning an uncommonly good salary.

At the top of the tree our lucky officer reaches £6 a day as Admiral of the Fleet, but as on having reached this rank he is probably past the age of employment, his actual income (half-pay) will be about £1,000 a year. The energies required to arrive at this rank are such that similarly in private life the income would probably result in something like £50,000 a year.

Taking things all in all and right away through it may be said without exaggeration that the financial prospects of a naval officer and those of a clerk in an ordinary business house, are about one and the same. As a money-making profession the Navy is useless to any man. That is why it remains what the Labour Party often describe it as the "preserve of the upper classes."

The Navy is a glorious profession, but there is certainly "no money in it."

Recently, in response to a public demand for "democratising " the Navy, it has been made possible for an ordinary bluejacket to reach the highest ranks. After passing certain examinations, he can become an acting Mate, and thereafter Mate, and so on to sub-lieutenant and upwards. To assist him in his career, he gets a certain amount of additional pay enough to enable him to live in some fashion without private means. Here, however, the sacrifices are great, and the situation from a financial point of view very poor. The nation's general attitude in the matter is that being a naval officer is a situation of such social importance that bare minimum of pay should be thankfully received.

The life of a naval officer is a strenuous one. He is cut away from home ties except at irregular intervals, and the responsibility on him is enormous. For example: a lieutenant on duty as watch-keeper, having the control of a two-million pound battleship, on which the issues of this war between England and Germany rests, by the time he has paid unavoidable expenses, for uniform, subscriptions, etc., his receipts are just about the same as a driver on the London General Omnibus Company, who drives his motor-bus through the streets of London. Whatever his rank that is about the utmost clear earning that any naval officer can hope for!

There are a variety of other officers in the Navy known as "civilian officers," in distinction from those which I have described, who are known as executive officers. The executive officer is distinguished by a curl on the top stripe of his uniform. These " civilian " officers who are non- combatants, although they have to take just the same risks in battle, are in a general way paid a little better than the other naval officers, but not very much, while their chances of the "glory part" of the business are infinitely less. The branches are:

Engineers. These are distinguished by having purple between the gold stripes on their arms. In case of an action their duties are below, where they run little risk of being injured unless the ship should be badly knocked about or sunk. In that case death is their inevitable fate generally a very unpleasant death, by being more or less boiled before they are drowned.

The normal duties of an engineer are obvious, so need not be detailed. On his efficiency depends whether the ship can carry through the duties assigned to her by the Admiral.

At the present time engineer officers are entered in exactly the same way as executive officers, though they used to be entered separately.

Paymasters (white between the stripes). In the piping times of peace, looking after the clerical work of the Navy, arranging for pay in advance to the men, and matters of that sort are their duties. In case of war their duties are to assist doctors.

Doctors wear red stripes on their arms and are officially known as surgeons. Every large ship carries two, small ships one. Little ships, like destroyers, do not carry any at all ; should they require a doctor they have to depend on a large ship being able to supply one.

In times of peace the duties of a doctor are not very onerous ; in times of war doctors are liable at any moment to have to put in a couple of day's duty or more without rest.

Their station is somewhere down in the bowels of the ship; their fate if the ship goes under is certain death.

Accountant officers, that is to say, paymasters, are entered as "clerks" by an entirely separate examination. So also are doctors, who enter at about the age of 24.

In addition, all large ships carry a chaplain of the Church of England. A naval man is allowed four religions, Church of England, Roman Catholic, Presbyterian, or Wesleyan. To one or other of these he must conform. If he has no particular choice when joining, he is entered as Church of England and is ministered to by the chaplain. Otherwise when the ship is in harbour he is, so far as possible, allowed to go ashore and attend whichever of the other three denominations he has announced himself as belonging to on joining. Whatever religion he joins on the chaplain is generally known afloat as the "Padre," or "sky pilot." He occasionally acts as Naval Instructor also. Naval instructors wear blue between the stripes and have to teach midshipmen and other budding naval officers various scientific matters connected with their profession.

In war time chaplains and naval instructors are detailed to assist the doctors.

All officers in the Navy have to provide their own uniforms, but all receive a small sum from the Government for messing. In a regiment of soldiers all officers mess together. In the Navy, on the other hand, a captain lives in solitary state by himself, and has his own table and apartments. The other officers, that is to say, commanders and lieutenants and civilian branches of corresponding rank mess in the "ward room" and live in cabins. The junior officers, i.e., sub-lieutenants, midshipmen and corresponding civilian ranks, live in the "gunroom." Some of the more senior of them have cabins, but most of them sleep in hammocks, the same as the men do.

Elsewhere in a ship there is another miniature wardroom the warrant officers' mess where the W.O.'s live and have their being. These warrant officers are those who have risen from the lower deck by methods described in the next chapter. Their various ranks are Warrant Officer and Chief Warrant Officer, this last being a commissioned officer. After very long service they achieve promotion to lieutenant.

But by this time they are always far too old for ordinary employment afloat. They get appointed to " shore billets," and so gradually pass on to retirement.

It is only within the last two years that any bluejacket has had a chance of becoming a real officer on the effective list.

Theoretically this boon is immense. In practice the pay of a naval officer is so small that it is next to impossible to live on it.

There are other officers not yet mentioned. I refer first to the Royal Naval Reserve known for short as "R.N.R." These are officers of the Merchant Service who, for a trivial bit of pay have put in a year's service afloat with the Fleet.

In addition thereunto are the "Supplementary Lieutenants." These are officers of the Mercantile Marine who have taken on regular naval jobs. Their reward is in the ordinary way a retiring allowance which is very nearly half what they might have looked for had they stuck to the Merchant Service.

Also there are the R.N.V.R. These are wealthy yachtsmen and folk of that sort. They are volunteers pure and simple. The "idle rich," is, I believe, their correct technical tally with the mass of folk in these islands.

"D... d nuisances" is nearer the official term for them. "Black legs" is the term applied to them by the merchant seamen of the R.N.R.

They are a lot of "silly enthusiasts" outside the scheme of everything an entirely British product.

When this present war started it is safe to say that they were far more dangerous in aspirations than in fact. But they were most desperately in earnest, and just at the present time men desperately desirous of getting some German blood are likely to get it somehow sooner or later. This particular force is our "last reserve." Properly organised, it is capable of indefinite expansion, for it consists of men familiar with every nook and cranny of our coasts.

IX. PERSONNEL: MEN.

There are several different branches of men in the Royal Navy, the two principal being bluejackets and stokers.

Bluejackets commence their career as boys at the age of 15. They are drafted to Shotley Barracks and thence to various training establishments, where they undergo elementary courses in gunnery, torpedo, signalling, navigation, etc. Their pay commences at sixpence a day.

After a time they are rated Ordinary Seamen at is. 3d. a day, then onwards to Able Seaman at is. 8d., Leading Seamen at 2s. 2d., Petty Officer at 33., and Chief Petty Officer at 33. 8d. In all cases a man can qualify himself for extra pay, though not to any very great amount.

Food for the men is provided by the Government. It is uniformly good and moderately plentiful. In addition there are allowances of rum (or money in lieu thereof), also an allowance of tobacco.

On board every ship there is a canteen where additional luxuries can be purchased, to supplement the regular Navy ration.

On joining, a certain amount of uniform is provided, but otherwise throughout his career a man has to find his own. The cost of this to the men varies, as those who are anxious to get promoted generally indulge in smart uniforms, which they make themselves, or get done by sailors who specialise in tailoring which afloat is known as "jewing."

In addition, all the above bluejackets enter for ten years' consecutive service, after which they can leave if they like and go into the Royal Fleet Reserve. Or they can sign on for a further period of twelve years, which entitles them to a pension which depends on their rank and character on leaving the service.

In addition, however, there are a good many men entered direct for five years' service. These men, who are nicknamed " ticklers," are put to do the less important work on shipboard, and have no prospects of promotion.

A Petty Officer, if he has done well, can qualify for and become a Warrant Officer. His pay is then from 6s. a day upwards, with allowances in addition.

A Warrant Officer is equivalent to a Sergeant-Major in the Army. They are generally known as the "backbone of the Service." Their duties are many and onerous. The old familiar branches of gunner, bo'sun and carpenter still exist.

The stoker branch are not entered as boys, but come along as young men. Other- wise their career is as stated previously.

There are a variety of other branches, as for example, joiners, shipwrights, blacksmiths, plumbers, painters, coopers, armourers, electricians, sick bay men, writers, cooks, ship's police, bandsmen, servants, stewards, tailors, shoemakers, etc., and there is even a rating known officially as " tindal of seedies."

The men of the Royal Naval Reserve (R.N.R.) are merchant service sailors who qualify with a certain amount of annual drill.

The Naval Volunteers were dealt with in a previous chapter. It only remains to add that the bulk of the "ordinary seamen" in it are rather better off financially than the ordinary captain, R.N.

X. THE ROYAL MARINES.

The simplest non-technical explanation of the Royal Marines is that they are "soldiers under Admiralty orders." This is not by any means an exact definition, as a marine when afloat does bluejacket duty under his own officers, manning guns, etc., while ashore his duties are of a military nature.

The marines make up, roughly, a quarter of the ship's company. They were originally instituted as a special police force in days when sailors were very unruly, and it is their proud tradition that when mutinies took place in the past the marines have always proved faithful.

The official motto of the marines is "per Mare, per Terram," and some idea of the services performed by them in the past may be gleaned from the fact that it was once proposed to give the " Sea Regiment " colours, such as the soldiers possess, with the names of their various battles emblazoned. When the scheme came to be gone into, however, it was discovered that no flag would contain all the necessary names!

There are two branches of Marines: Artillery, always known as the Blue Marines, and the light infantry, known as the Red Marines. They are blue or red according to the colours of their tunics.

Marine officers are entered by examination in the same way that army officers are entered, and all their ranks are military titles. The commencing pay of an officer as second lieutenant is £3.3d. a day. A colonel receives about £400 a year. There are, of course, various allowances which raise the pay a little.

The men enter as privates at is. 4d. a day. Unlike the bluejackets, the marines get their uniforms provided by the Government.

There is no short service in the Marines such as exists in the army, all being long- service men. The bulk of them at any one time are what is known as "old soldiers."

XI. HOW A FLEET GOES INTO ACTION.

In the old days of a hundred years ago, when ships depended on the sail for propulsion, the root idea of successful battle was "cutting the line."

Nelson employed it at Trafalgar not as a new device, but as an old and well-tried one. "Cutting the line " worked somewhat as follows:

There were two things for which every admiral manoeuvred. Of these the first was the "weather gauge." In the old sailing days a ship under sail naturally heeled over to the breeze, and could not use her lower tier of guns. Consequently the ship (or fleet) which got the weather gauge, it was able to use both tiers of guns, had an immense advantage over the enemy.

The other old advantage was "cutting the line." This meant sailing down towards the enemy; cutting some of them off and hammering them out, long before the rest could beat about and return.

All this sort of thing is, of course, quite dead nowadays. But the integral principle is just the same to-day, and will ever remain the same. It consists in manoeuvring for an advantage.

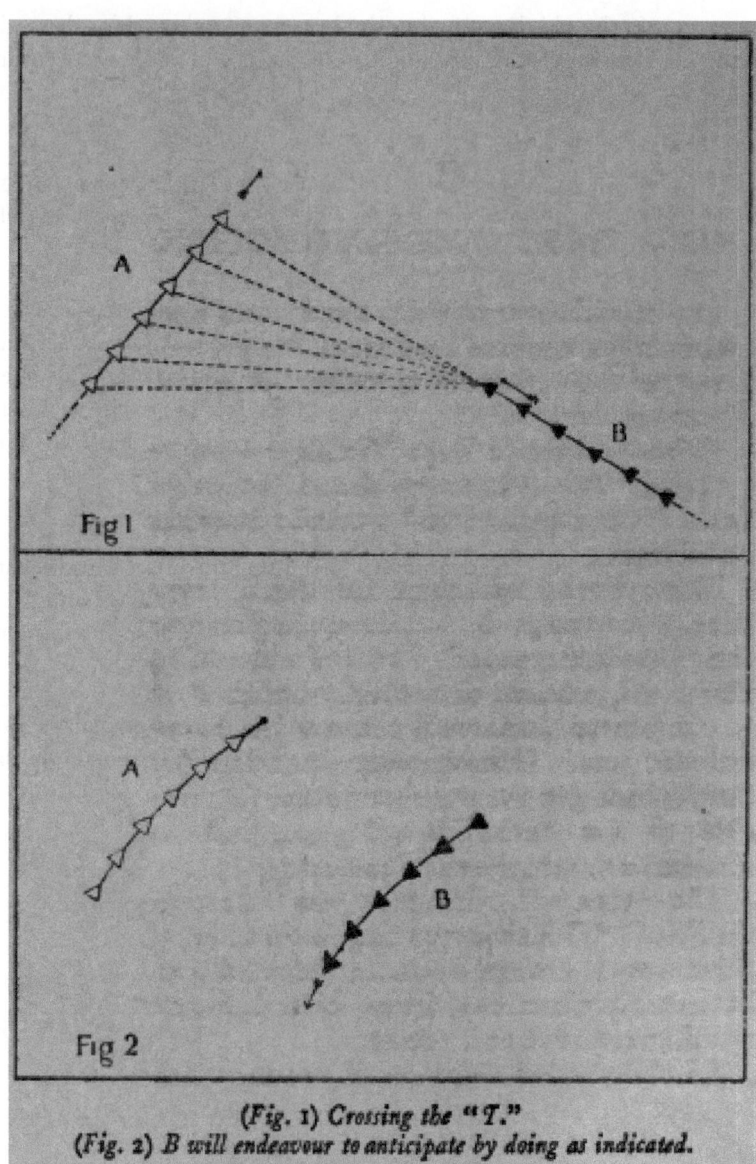

(Fig. 1) *Crossing the "T."*
(Fig. 2) *B will endeavour to anticipate by doing as indicated.*

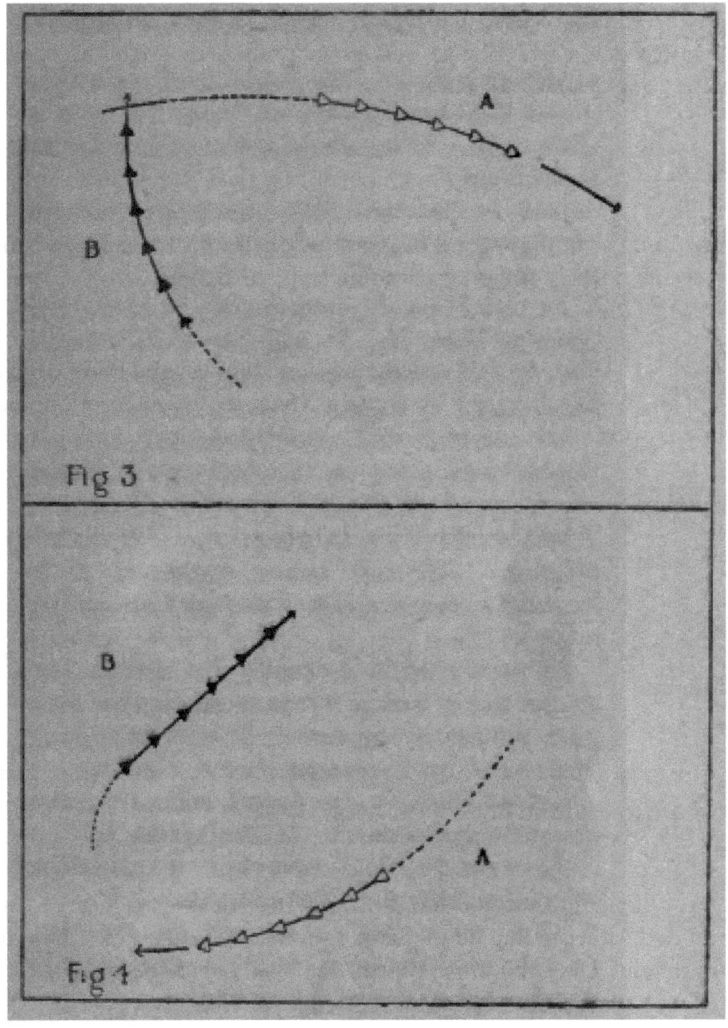

(Fig. 3) An endeavour to cross the "T" on the tail of A.

(Fig. 4) While A endeavours to do the same by B.

Just at present that advantage rests in managing to accomplish a manoeuvre known as "crossing the T."

In this Fleet A manages to get across the bows of Fleet B. It will be noted that all the A ships can concentrate their fire on the leading B ships. (See diagram 1.)

Of course B will not charge into A, as it would have done in the old days. Steam has reversed all the old formulae of victory. B will meet A by turning as in the second diagram. All the same, however, A is bound to secure a heavy fire on the leading ships of B.

In substance it is exactly the same thing as the thing which happens when two men play draughts together. B will turn long before the first position seems possible.

He will seek to get round behind A, and reverse the situation. (See diagram 3.)

A, of course, will endeavour to repeat the original situation. (See diagram 4.)

This can either go on and on for ever, or until one admiral will manage to gain and hold the advantage point.

Failing this there is nothing left but for the two fleets to steam past each other in line till the better gunners have hammered the more indifferent. Thus and thus only are Trafalgars made nowadays.

If we care to examine history we shall find that in Nelson's days the general principle was exactly the same.

A modern fleet-action where the battleships are concerned must necessarily be short, though the manoeuvres indicated may last for a very considerable time.

It would probably be heralded by a good deal of desultory fighting between the scouting cruisers on either side.

After it was over the destroyers of the victor would be hotly engaged in endeavouring to sink what was left of the vanquished, also likely enough in isolated combats with hostile destroyers.

No navy can afford to have an indecisive fleet action. It is necessary to have it absolutely conclusive, so that it means the complete annihilation of one side or the other.

XII. GUNS.

The gun, as we understand it, first came into use after armour was invented. Till armour came along, penetrative power mattered very little. The object was merely to hit the enemy with a smooth bore cannon which first fired cannon-balls and later shells. It was to defeat the shells that armour came to be applied to ships.

Once, however, it became the custom to clothe ships with an "impenetrable coat of mail " the entire situation was altered ; the ideal of every gun-maker was to penetrate that coat of mail.

To this and this alone may be attributed the extraordinary advances in gun power which this generation has seen. It began with oval and hexagonal shot, and thence proceeded with the introduction of twisted grooves, or "rifling." This allowed of the use of pointed projectiles, which were first supplied with studs to fit the grooving. Nowadays special bands which adapt themselves to the rifling are substituted. The object of rifling was to secure increased range.

The battle guns versus armour, was a long one. No sooner was a gun invented capable of penetrating armour than a new and superior kind of armour was invented, capable of keeping that projectile out of a ship's vitals. Eventually, however, the gun forged ahead.

On the "proving ground" where guns are tested against armour plates the gun won hands down. It is easier to increase the size of a gun than to increase the thickness of armour, the relative increases of weight concerned being totally disproportionate.

On the other hand battle experiences of ten years ago between Russia and Japan by no means indicate that, so far as armour is concerned, defence is knocked out by the attack. Roughly, the contention under which we have gone to war is that under practical conditions the gun will not penetrate armour as expected, because projectiles are never likely to hit exactly dead true, and all British armoured warships are armoured along such lines. The Germans, on the other hand, have sought after protection more than we have, adopting the other of the two rival theories :

(1) That defensive sacrifices should be made in order to inflict greater damage on the enemy.

(2) That to inflict damage on the enemy it is necessary to be well protected against his return fire.

It is unwise to lay down any generalisation as to which theory is the better. One can merely indicate the different lines along which British and German naval policies have proceeded.

The rate of fire is an important thing with guns. In this connection the normal possible rates are about as follows:

15	inch	1.2	shots per minute
13.5	"	1.5	"
12	"	2	"
9.2	"	4	"
7	"	6	"
6	"	10	"
4.7	"	12	"
4	"	15	"
12 pounders	"	25	"

Which, of course, works out that the bigger the gun the slower it fires.

Here one may explain that a 15 -inch gun is one whose bore is of 15 inch in diameter, and so on down.

The 12-pounder is of 3-inch bore, but for some reason guns of this size in the British Navy are spoken of as "12-pounders," because they fire a shot of 12 pounds weight. In all other navies they and lesser guns still are always spoken of in "bores."

In the old British Navy it used to be the custom to speak of guns as "32-pounders," "64-pounders," "100-pounders," and so forth. This continued until 300-pounders were arrived at.

Then a system was introduced of speaking of guns by weight for example an " 8-ton gun." These were muzzle-loaders.

When breech-loaders came in, for purposes of distinction, it became the custom to speak of them as " 12-inch," etc.

For similar reasons, when the Hotchkiss anti-torpedo boat guns first appeared with metric designations, such as 37 m/m, 47 m/m 52 m/m and what not, it was deemed inconvenient to refer to them in inch equivalents, some of which, such as "2.2," or "1.4," were confusing, and a Nelsonian notation by the weight of projectiles was adopted.

But everything which was a "ship gun," i.e., carried by any ship as part of her main armament as distinguished from her anti- torpedo attack armament has continued to be alluded to in terms of inches.

In so far as "pounders " could be applied, the 4-inch gun is a 25 - pounder (or occasionally a 32-pounder). The 4.7-inch is somewhere about 40 pounds. The 6-inch is in the neighbourhood of 100 pounds, the 7.5 fires a shot of 200 pounds. The shot of 9.2 weighs 380 pounds, while that of the 12-inch weighs 850 pounds.

The projectile of the 13.5 was originally of 1,250 pounds, but in later models rose to 1,440 pounds. The exact weight of the 1 5 -inch projectile has never been published, but it is probably not far short of a ton.

An earlier chapter explains why the big gun, which fires slowly with a heavy projectile, is of more use to the Navy than the lesser gun which fires a lighter projectile.

Of all the guns mentioned the most accurate for short ranges is the 4-inch. The 4-inch of modern times is not to be confounded with those 4-inch guns of the Pegasus, which could not reach the German cruiser which destroyed it. The guns of the Pegasus w r ere very old pieces, which were quite incapable of ranging to any distance which matters in modern naval warfare.

The modern 4-inch, on the other hand, is an entirely up-to-date gun, with which it is next to impossible to fail to hit. It is the gun of our destroyers, and is also mounted in a number of our more modern light cruisers.

The rapidity of firing from guns is greatly governed by the fact as to whether the ammunition for them can be man-handled, or must be worked by machinery.

The limit in this respect is the 6-inch gun, with its loo-pound projectile. Experience has indicated that beyond that weight it is not possible for men to hand along the projectiles for any length of time.

In some of their ships the Germans tried to utilise a heavier gun known to us as the 6.7, which fires a projectile of 132 pounds weight, but they presently learned that this was too heavy for a single man to manipulate for any length of time, and so in their

Dreadnoughts they also reverted to the 6-inch gun with the projectile averaging a little over 100 pounds.

The principal feature of a gun, in the matter of straight shooting is what is known as its initial velocity that is to say, the rate per second at which it leaves the muzzle. The faster its velocity the less it has to be sighted for and the more certain it is to hit the target. Here again, reference is made to the previous diagram. If a projectile be light it cannot long keep up its initial velocity.

When the British cruiser Pegasus was engaged with the German cruiser which sank her, her old guns had to be elevated to the utmost angle in order to obtain any chance whatever of hitting the enemy. The German cruiser, on the other hand, had merely to fire at something like point-blank range, because the velocity of the guns was so much greater and little elevation was required.

Any modern 12-inch gun is theoretically at any rate as capable of getting through hostile armour as a larger weapon. But at long range it is less accurate than the bigger pieces. Consequently, every Navy is going in for bigger and bigger guns on the principle that the bigger the gun the easier it is to secure a hit.

I have referred to the rate of fire. But there is another factor, and that is the weight of projectiles from one discharge, and the distance at which these projectiles are effective. Roughly, the distances are as follows:

12   inch guns up to 5 miles

13.5   6 miles

15   7 ½

At eight miles off a 15 -inch, if it hits square, will (theoretically) get through 9 inches of the latest type of armour.

We come now to the weight of the projectiles discharged by a single broadside. They are as follows:

8-12 inch from the Dreadnought 6,800 Lbs.

10-12 inch from Neptune 8,500 Lbs.

10-13.5 inch from Orion 12,500 Lbs.

8-15 inch from Queen Elizabeth 15,600 Lbs.

The enormous increase of weight discharged explains how it is that battleships so quickly get out of date and have to be replaced by later productions. It is the march of invention which does the mischief and costs the money. Nelson's Victory was 50 years old at the battle of Trafalgar, but as there was no "march of invention" in those days she was just as efficient as any brand-new ship. To-day a five-year-old is already at a considerable disadvantage.

XIII. TORPEDOES.

A torpedo is best described as a small uninhabited submarine. The original inventor of the automobile torpedo was an Englishman named Whitehead, who, meeting with no appreciation in his own country, established works in Fiume, in Austria. His early efforts were, of course, very simple and trivial things they might travel a hundred yards, or they might not. And they also had an unpleasant habit of "doing the boomerang," i.e., coming back at the boat which fired them.

Long ago, however, all these defects were overcome, and a torpedo can now travel up to five miles at a considerable speed.

Till comparatively recently all torpedoes were of 17.7 inches in diameter generally known as the "18-inch." To-day, however, this 1 8-inch is superseded in all modern vessels (except of the French Navy) with a 21 -inch torpedo which, being larger, carries a far heavier explosive charge. Details as to the charge are more or less confidential, and wisely so.

Not, however, that wisdom has always been a salient feature in connection with torpedo matters. An absurd situation arose many years ago when the perfected Whitehead first appeared. The perfection was arrived at in a part of the torpedo known as the " balance chamber "the integral part of the whole mechanism.

When any Power purchased the right to utilise Whiteheads the secret of this balance chamber was only revealed under oath of secrecy to certain torpedo lieutenants a sort of Masonic secret. Even an admiral might know nothing about it, lest by an incautious remark the possible enemy might find out something.

All of which was very well as a precaution, but it presently came about that every navy in the world was thus hiding secrets known to every other navy!

To-day, of course, there is no particular secret ; and there are other makes of torpedoes the Schwartzkopf, in Germany, the Bliss-

Leavitt, in the United States. Neither apparently is quite so good as the Whitehead, but they are amply good enough for all practical purposes. And "practical purposes "is" the thing that matters."

Whatever the design all torpedoes are constructed on the same general principle. This principle is as follows:

(1) Head of the torpedo in which the explosive charge is carried. This is exploded by means of a "pistol." The " pistol " mainly consists of a striker which, on hitting anything, hits the detonator a small amount of fulminate of mercury, which ignites and explodes the explosive charge.

(2) The "air chamber." This contains compressed air which supplies the motive power of the torpedo. In modern torpedoes there are various devices for super-heating this air, and so securing extra efficiency.

(3) The "balance chamber." This regulates the depth at which the torpedo will run. It can be set for any depth. The general principle is that should the torpedo get higher than its depth, the water comes in and makes it sink a bit by altering the alignment of the horizontal rudders. If, on the other hand, the torpedo is too deep down a series of springs incline the rudders until they automatically make it come up again.

(4) The engines. These are worked from the compressed air carried in section

(5) The "buoyancy chamber." This is simply a chamber to give the necessary buoyancy to the torpedo.

(6) The "tail." This consists of two propellers, whereby the torpedo is driven, and also the rudders already referred to.

In addition to this all torpedoes are now fitted with a gyroscope, a heavy revolving wheel which tends to prevent anything which might otherwise deflect the torpedo from its original line.

Various other additions to torpedoes have been patented from time to time. For example, there is a device whereby, should a torpedo miss a ship by getting astern of her, the wash of their propellers will cause the torpedo to circle round and re-attack from the further side.

There is also an American invention which perhaps is really being used in the German Navy at the present time. The principle of this is to substitute a short 8-inch gun for the explosive head. When the torpedo makes contact, this gun fires a high-explosive shell right into the vitals of the ship struck.

Early torpedoes used to cost about £500 each. To-day the cost may well run up to nearer £2,000.

Torpedoes are discharged from a species of gun known as a "torpedo tube." These are of two kinds those carried above water in small vessels, and those fitted under water in big ships (submerged tubes).

The general principle of both is the same. A small charge puts the torpedo into the water thereafter, it progresses by its own motive power. The only essential differences are that whereas the above-water tube merely plumps the torpedo into the water, the submerged tube is fitted with a. bar which guides the torpedo clear of the ship's side. Else it might be deflected by the motion of the ship firing it.

The other difference is that the above-water tube can be trained like a gun on the enemy, whereas the submerged tube is fixed, and the ship discharging must be manoeuvred so that the tube can bear.

This entails loss of speed and position to use the submerged tube. On the other hand, should a shell burst on or near an above-water tube the odds are that the torpedo explodes and destroys the ship carrying it.

Some six months before the present war a French naval officer proposed that tubes should be carried above water in armoured turrets and used as guns are used. But the war happened long before this suggestion could be tested and made use of if it turned out to be practical.

We should be careful not to assume that ships are defenceless against torpedoes despite the fact that (according to German reports) a single submarine sank three old cruisers of ours of the Cressy class.

Modern capital ships are armoured under water against torpedoes. The efficiency of this armour is a matter of theory, but there is as yet no reason to suppose that one or even two torpedoes would do more than temporarily disable a Dreadnought.

The defensive principle is based on the idea that internal under-water armour, somewhat set back, will confine the effects of a torpedo into comparatively narrow limits.

There is also "net defence," otherwise known as "Bullivant," after the firm which specialised in this direction. "Net defence " consists in a series of nets hung out from the attacked ship. These nets are composed of a series of steel rings a sort of steel fishing net hung out round a ship. Their main defect is that they greatly reduce the speed of a ship. Also every modern torpedo is fitted with automatic nippers to work its way through the nets.

But there is yet no reason to suppose that such " net cutters " will be effective against a ship in motion under war conditions.

For many a year the British Navy was the only one which trusted to "net defence." Just before the present war the British Navy suddenly gave up net defence, while the German Navy just as suddenly adopted it.

From which the assumption is that the Germans had become possessed of some appliance against which nets were of no avail; but that we have nothing of the same sort or are supposed not to have. Probably the Germans are using the sub-marine gun referred to previously.

GLOSSARY OF NAVAL TERMS.

Abaft, the end towards a ship's stern.

Abeam, in line with the middle of the ship. About the same thing as "amidships."

Admiralty, the Board which controls the Navy. The Administrator of the Navy. The "First Lord" is a Cabinet Minister. The " Sea Lords " act as advisers to him.

Adrift, term used by bluejackets who have broken leave.

Aerial, the essential wires whereby wireless messages are received or sent.

Aft, the stern : back end of the ship.

Amidships, the middle of the ship.

Andrew Miller (occasionally Merry Andrew), nautical nickname for the British Navy.

Armour, specially treated steel plates which are placed on the sides of the ship and its turrets to keep out projectiles.

Armour Belt, (see Belt).

Armour-piercing Shell, shell with a strong point intended to penetrate armour and then burst.

Barbette, a fixed armoured tower inside which guns revolve. Design of French origin as opposed to the contemporary British idea of a revolving turret.

Guns in all battleships are now mounted in barbettes, but the term " barbette " has fallen into disuse, the word "turret" being used instead.

Barque, a three-masted vessel with square sails on the fore and mainmast, but not on the mizzen.

Barquetine, a three-masted vessel with square sails on the fore mast only.

"Bailment de Guerre" French term for any warship. Vaisseau is never used for a warship in France, except to describe an old three-decker of about a hundred years old.

Battery, this term is usually applied to inferior guns which are mounted either to support the big guns or for use against torpedo attack. "Auxiliary battery " is the term now applied.

Beach, a naval expression indicating going ashore, e.g., "going to the Beach."

Beam, the breadth of a ship. "Beam " is the measurement of the maximum breadth.

Bells, time at sea is indicated by bells which are struck to indicate each half-hour of the watch. For example, noon is eight bells, 12.30 one bell, 1 o'clock two bells, 1.30 three bells, 2 o'clock four bells, and so on until 4 o'clock, which is eight bells. The watch is then changed and the bells begin again with one bell for 4.30, etc.

Belt, the portion of armour protecting a ship's water line.

Bilge, the bottom of the inside of a ship.

Bilge keels, special keels projecting from the bottom of a ship to minimise the rolling.

Blockade, when a hostile fleet is off a harbour or coast in sufficient strength to prevent ingress or egress of anything, it is said to have established what is known as an "effective blockade."

Blockade-runner, a merchant ship which takes chances of being able to run into a blockaded port without interception by the blockading force.

Bluejacket, term applied to any sailor whose duties are entirely on deck.

Blue Water School, the name applied to those who hold that shore defences are mostly unnecessary, and that the defence of the coast should rest entirely with the fleet.

Boat, destroyers and torpedo boats are frequently spoken of as "boats." A large ship is never spoken of as a " boat " in the British Navy.

Boilers, these are, of course, the steam generators. Various types exist, but in the British Navy all are either Yarrow (with tubes going at an angle of 45 degrees into a drum), or Babcock (with the tubes placed more or less horizontally). In either case these contain water which passes through the furnace fire and is heated accordingly, with the result that steam is generated at the other end. Some of the older ships have Belleville boilers, also tubes inclined slightly out of the horizontal.

Boom defence, this is a series of heavy spars or masts, lashed together to old vessels in order to block a harbour's mouth. A boom defence has usually one entry which can be opened or closed at will in order to admit ships. The front part of a ship.

Bridge, part of the ship from which the vessel is navigated in peace time or when not in action.

Brig, a two-masted vessel with square sails on both masts.

Brigantine, a cross between a brig and topsail schooner.

Broadside (1) the side of a ship. (2) the term is also used to indicate the whole of the available guns on one side of the ship firing.

Bulkhead, partitions in the inside of a ship. These exist to strengthen the hull and to localise the entry of water should the outer skin be pierced.

Bunkers, places along the side of a ship, up or below the water-line in which coal is stored. So long as a bunker is un-emptied this coal is a valuable asset in stopping or minimising shells which have penetrated the armour.

"Buzz" nautical slang term for any rumour which obtains partial credence.

Cable, (1) the chain to which the anchor is attached. (2) a nautical measurement = 200 yards.

Capacity, this means the amount of coal and oil which a ship can carry. There are two sorts of capacity: (1) Normal, which is what a ship can carry on her designed displacement. (2) Full, the amount she can manage to carry, and which she always does carry.

Capstan, the mechanism by which the anchor is hoisted up.

Casemate, a fixed armoured structure inside which a secondary gun (that is to say, an 8 or 6-inch gun) is worked. As a rule the front is six inches thick and the rear two inches.

Charlie B., Naval name for Lord Charles Beresford.

Chartbouse, an erection on the bridge in which the charts are kept by which the vessel's course is steered.

"Chief," senior Engineer Officer of any ship.

"C.O.C.," Commander-in-Chief.

Complement, the number of officers and men forming the ship's crew.

Conning tower, a heavily armoured structure in the forepart of the ship, either under or immediately in front of the bridge. It is the post of the captain and others responsible for the steering and control of the ship in action.

Cowl, a ventilator.

"Der Tag " (" the Day ") a toast among German Naval officers meaning the time that they expected to fight and beat the British Navy and establish Germany as the world's premier navy.

Destroyer, a small vessel fitted with torpedo tubes intended to attack big ships. Originally designed to attack smaller vessels of her own sort.

Division, a group of ships three or four in the case of battleships, and anything up to twenty in the case of destroyers, which act together.

Doctor, in the merchant service is the slang term for the ship's cook ; but it is not used in this connection in the Navy.

Double bottom, a system of sub-dividing the bottom of a ship into a number of small compartments with an inner and outer skin. It is designed to minimise injury to the ship's bottom.

Draught, in the ordinary way the draught of a ship is her " mean draught," that is to say, the difference between the amount of immersion forward and the amount of immersion aft, which is generally several feet more. The "maximum draught" indicates the least depth of water in which a ship can travel.

Ebb tide, the tide going out, i.e., falling.

End-on, a ship with her bow or stern to the enemy is " end-on."

Ends, the "ends " of a ship are either the bow or the stern. In some designs they are heavily armoured; in others they are left entirely unarmoured. Hence the expression "unarmoured ends."

Even keel, when there is little or no difference between the fore and aft draught of a ship as designed, she is said to be of "even keel."

Fairway, the centre of a channel or harbour.

Fathom, a nautical measure of depth = six feet.

Fighting tops, a fighting top is anything on a mast from which guns can be fired. Nelson was killed at Trafalgar by a man who fired on him from aloft. Twenty years ago every warship carried a certain number of light guns on her masts. To-day the mast being required for fire control stations and searchlights no guns are carried above the upper deck.

Fire controls, stations in the masthead from which officers can take the range of an enemy and having passed this down below, the guns know whether the shots are short or too far beyond. The range used down below is corrected accordingly.

Flatfoot, an old term whereby bluejackets described themselves.

Fleet, a number of squadrons collected together.

Flood tide, the tide corning in, i.e., rising.

Flying topmast (see Topmast).

Forecastle (Foke's'le), the forward part of a ship. Usually inhabited by the crews, but in the earlier Dreadnoughts the living space of the officers.

Foremast, the first mast of a vessel counting from the bow.

204

"Gobby," Naval term for a coastguard. Hence reserve ships are known as "Gobby ships."

"Goeben," British naval term for a coward. Originated in August, 1914, on account of the show German battle-cruiser Goeben having run away from the British light cruiser Gloucester.

"G.Q" general quarters ; every man at his war station on board the ship.

"Gunnery Jack" the chief officer responsible for the guns and gunnery. Gunroom, a place in the after part of a ship allocated to junior officers and midshipmen.

Heavy weather, a sailor who chances to have got drunk describes it as having "made heavy weather."

Hoists, tube through which ammunition is passed up to the guns.

Hydraulic gear, big guns and their mechanism are far too heavy to be operated by hand. Consequently machinery has to be employed.

Hydraulic mechanism is the most popular, because being worked by water any leak or defect is easily detected. The principle is that "water always finds its own level."

Interned, under International Law, should a hostile warship enter a neutral harbour and not leave it within twenty-four hours she becomes "interned," i.e., she is taken possession of by the neutral until such time as the war is over.

"Jackie," Naval name for Admiral Lord Fisher.

"Jeune Ecole," name applied to a body of thinkers (mostly French), holding the idea that naval success is best to be secured by torpedo operations and attack on commerce.

Kaiser Bill, a naval expression for bluff that cannot be supported.

Knot, a nautical measure of speed 2,000 yards (a sea mile). A ship's speed is calculated by how many sea miles she can do in an hour, the result being expressed in knots. For example: 25 knots an hour = able to do 25 sea miles in an hour (roughly 28 land miles).

Leeward, (to), naval expression for having got the worst of anything. This expression dates from Nelson times.

Liquid fuel, term applied to any oil used for the propulsion of warships: (1) Used to increase the heat of the coal fire under the boilers, or (2) maybe the sole heat generator.

Log (1) a daily record of everything which takes place on shipboard.

(2) a ship's speed indicator which " logs " the velocity.

Lower deck, (1) the deck immediately above the water line.

(2) The term " lower Deck " is also used to indicate the ship's company, in the same way that officers are spoken of as the "Quarter Deck."

Magazines, the portions of a ship in which the ammunition is stored.

Mainmast, the second mast of a ship counting from the bow.

Matlo, name used to describe themselves by British bluejackets. Falling into disuse. Corruption of the French matelot. The term dates from the Crimean War, when the British and French sailors worked together.

Mizzen, the third mast of a ship counting from the bow.

Moan, nautical term for any complaint about things.

Monkey house, the top of the chart house, from which the officer on duty usually works the ship, torpedo nets are hung out from the side of a ship to catch torpedoes. The advantage of nets is that they save ships from being torpedoed; the disadvantage of them is that, if the nets be out the ship can only steam at a slow speed.

"Neucloid" nautical term for those employed in reserve ships. (See "Gobby.")

Padre, the chaplain on board a man-of-war.

Pay, paymaster.

Pole mast, a mast in one piece only.

Potted air, artificial ventilation, common in modern warships.

Prize, an enemy merchant ship captured by a warship becomes the property of the capturing nation and is subsequently sold. Until recently the crew of the capturing warship took the profits on the transaction, but nowadays in the British the profits are spread over the entire fleet, the idea being that all who have done something towards the capturing deserve their share just as much as those who actually effect it.

Protective deck, this was first introduced for cruisers not otherwise armoured, but for a long time it has also been applied to battleships as an additional protection behind the belt armour. The principle is armoured plates inclined at an angle of 45 degrees. In a "protected cruiser "two inches of armour at an angle like this is theoretically equal to four inches or more of vertical side armour. There is some doubt as to whether this is true; but absolute data are unavailable. In a battleship with an armour belt the idea of a protective deck is somewhat different. The principal idea is that should shells get through the belt and burst, the pieces would be unable to penetrate the protective deck.

Quarter, either side of the ship aft.

Quarter deck,   (1) the upper deck in the after part of the ship.

(2) As this part is reserved for officers these are generally known as "the Quarter Deck."

Rake, when, instead of being upright, masts are on the slant, they are said to have a "rake."

Ram, some twenty years ago every battleship carried a projecting spur, fitted under water at the extreme end of her bow. The idea was to punch a hole in the enemy's bottom. Nowadays, owing to the progress which has been made in guns and torpedoes the ram no longer remains a weapon of account afloat. In some form or other, however, it is generally retained in the same way that soldiers still retain the bayonet, the idea being that, just as in every war the theoretically useless bayonet is found a determining factor, so the ram may yet prove of utility in certain circumstances.

Rangefinder, an instrument for finding how far away the enemy is. It works by two optical images of the enemy coming into clear conjunction. When this conjunction is correct, the exact range can be ascertained.

Reciprocating engines, this term is applied to all older methods of steam propulsion in which power is obtained by means of pistons being forced up and down cylinders by steam power.

Schooner, a ship with fore and aft sails only. Sometimes these are fitted with square sails forward. They are then known as topsail schooners.

Screen, originally this consisted of nets slung between the guns of a batter)- so that if a shell struck one gun the pieces from it would be localised. To-day a screen generally consists of two inches of steel between guns which are side by side in a battery. The purpose is the same. The word "screen" is also applied to a thin plate of steel fitted in front of light guns. It is of no value whatever for protection, but it is calculated to give confidence to the crew of the gun behind it.

Scuppers, the side of the ship under the bulwarks or rails, for drainage of sea water which comes aboard.

Sea lawyer, term applied to any sailor who is over-interested in his exact legal rights.

Sea Lord, a Naval officer serving on the governing Board of Admiralty.

Sea mile, 2,000 yards. About 1 1/7th of a land mile. Naval distances are expressed in sea miles as "miles."

Searchlight, an electrically operated machine like an enormous bull's-eye lantern used for projecting light at night and detecting approaching vessels.

Shell, any projectile containing a bursting charge either of gunpowder or high explosive.

Shield, this is a metal shield which may be anything from one to six inches thick, placed over a gun not otherwise protected, or occasionally over gum which are armour protected also. It gives no absolute protection to the men operating the gun, but it does protect the mechanism of the gun. One objection to a shield is that it may arrest and cause to burst a shell which would otherwise pass clear.

Shove off, Naval expression for leaving any place or person.

Sky Pilot, an old name for a chaplain. Seldom heard nowadays.

"Snottie," Naval slang for a midshipman.

Soundings, the depth of water expressed in fathoms as measured by a sounding line.

Spotting, technical term used to explain watching where shots aimed at the target fall. "Spotting " is done from a position on one of the masts.

"Squaddy," a squadron commander in the Naval Air Service.

Squadron, a small group of warships not sufficiently numerous to be termed a "fleet."

Starboard, the right-hand side of a ship looking forward. A green light is carried on this side at night.

Stem, bow of a ship.

Stern, extreme end of a ship aft.

"Tar" term used by the British public to describe sailors. Quite unknown afloat.

Topmast, the masts of nearly all ships are made in either two or three pieces. The second of these is known as the topmast. The third is the topgallant mast, though in the Navy the topgallant mast is always spoken of as "flying topmast."

Torpedo tube, an instrument for discharging torpedoes.

"Torps," usual nickname of any officer responsible for the torpedoes and wireless. The official designation is " Lieut. (T)."

"Tug," a name applied in the Navy to anyone called Wilson.

Turbine, the most common method of the propulsion of warships to-day. It is difficult to describe in non-technical terms, but in general principle it is similar to a water-wheel with jets of steam taking the place of the water.

Turret, originally a turret was a heavily armoured revolving tower containing a couple of guns with its base protected by fixed side armour. (See Barbettes)

Weather side, the side of a ship facing the wind.

Windsail, an erection of sail cloth, intended to take the part of a ventilating cowl.

Windward, (to wind'ard of), Naval expression to indicate an advantage of any kind.

Wireless, wireless telegraphy is under the control of the torpedo lieutenant and strictly confidential. The British Navy has easily the best "wireless" in the world.

DID2437913

L - #0068 - 070219 - C0 - 234/156/11 - PB - DID2437913